Atomic Absorption and Emission Spectroscopy

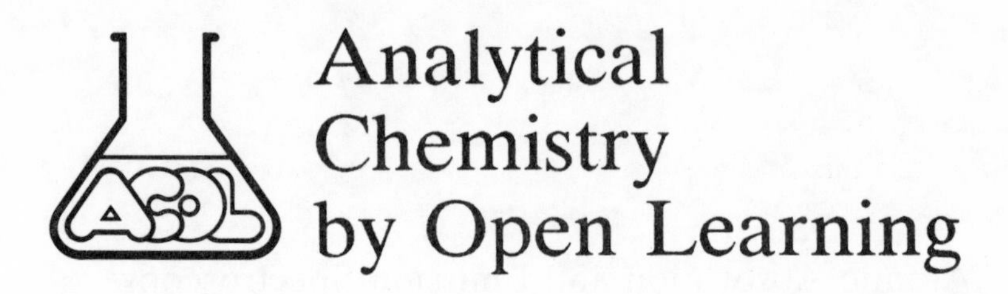
Analytical
Chemistry
by Open Learning

Titles in Series:

Samples and Standards
Sample Pretreatment and Separation
Classical Methods
Measurement, Statistics and Computation
Using Literature
Instrumentation
Chromatographic Separations
Gas Chromatography
High Performance Liquid Chromatography
Electrophoresis
Thin Layer Chromatography
Visible and Ultraviolet Spectroscopy
Fluorescence and Phosphorescence Spectroscopy
Infra Red Spectroscopy
Atomic Absorption and Emission Spectroscopy
Nuclear Magnetic Resonance Spectroscopy
X-ray Methods
Mass Spectrometry
Scanning Electron Microscopy and X-Ray Microanalysis
Principles of Electroanalytical Methods
Potentiometry and Ion Selective Electrodes
Polarography and Other Voltammetric Methods
Radiochemical Methods
Clinical Specimens
Diagnostic Enzymology
Quantitative Bioassay
Assessment and Control of Biochemical Methods
Thermal Methods
Microprocessor Applications

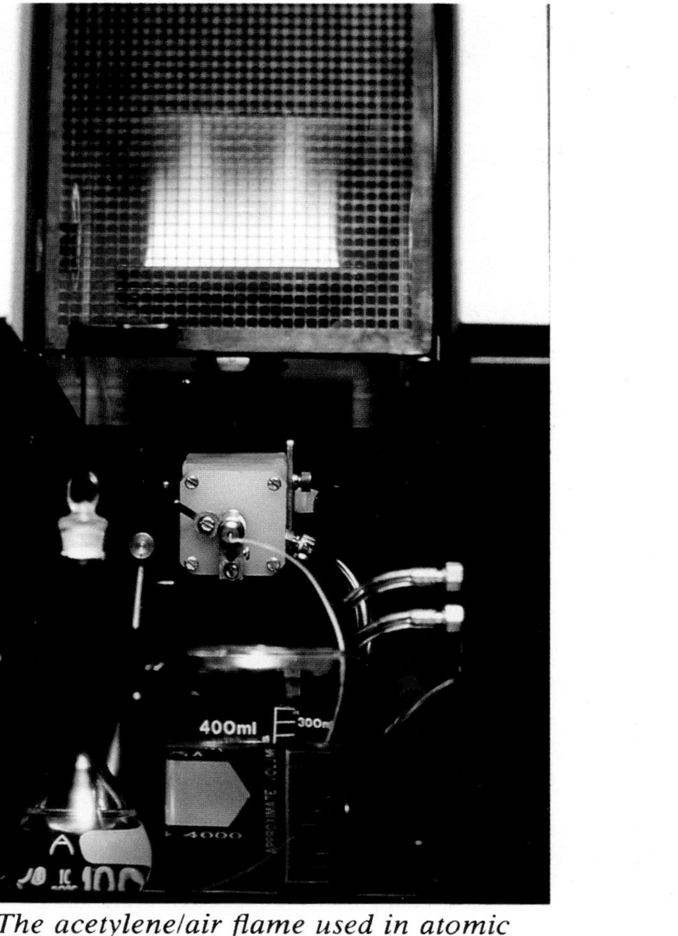

The acetylene/air flame used in atomic absorption analysis

Inductively-coupled plasma torch used in multi-element atomic emission-analysis

Atomic Absorption and Emission Spectroscopy

Analytical Chemistry by Open Learning

Author:
ED METCALFE
Thames Polytechnic

Editor:
F ELIZABETH PRICHARD

on behalf of ACOL

Published on behalf of ACOL, Thames Polytechnic, London
by
JOHN WILEY & SONS
Chichester • New York • Brisbane • Toronto • Singapore

Library of Congress Cataloging in Publication Data:

Metcalfe, Ed.
Atomic absorption and emission spectroscopy.
(Analytical Chemistry by Open Learning)
Bibliography: p.
1. Atomic spectroscopy. 2. Emission spectroscopy.
I. Pritchard, F. Elizabeth (Florence Elizabeth) II. Title.
III. Series: Analytical Chemistry by Open Learning (Series)

QC454.A8M48 1987 535.8'4 87-13378

ISBN 0 471 91384 7
ISBN 0 471 91385 5 (Pbk.)

British Library Cataloguing in Publication Data:

Metcalfe, Ed
Atomic absorption and emission spectroscopy.—(Analytical chemistry).
1. Atomic spectra
I. Title II. Prichard, F. Elizabeth
III. ACOL IV. Series

535.8'4 WC454.A8

ISBN 0 471 91384 7
ISBN 0 471 91385 5 Pbk

Printed and bound in Great Britain

Analytical Chemistry

This series of texts is a result of an initiative by the Committee of Heads of Polytechnic Chemistry Departments in the United Kingdom. A project team based at Thames Polytechnic using funds available from the Manpower Services Commission 'Open Tech' Project has organised and managed the development of the material suitable for use by 'Distance Learners'. The contents of the various units have been identified, planned and written almost exclusively by groups of polytechnic staff, who are both expert in the subject area and are currently teaching in analytical chemistry.

The texts are for those interested in the basics of analytical chemistry and instrumental techniques who wish to study in a more flexible way than traditional institute attendance or to augment such attendance. A series of these units may be used by those undertaking courses leading to BTEC (levels IV and V), Royal Society of Chemistry (Certificates of Applied Chemistry) or other qualifications. The level is thus that of Senior Technician.

It is emphasised however that whilst the theoretical aspects of analytical chemistry can be studied in this way there is no substitute for the laboratory to learn the associated practical skills. In the U.K. there are nominated Polytechnics, Colleges and other Institutions who offer tutorial and practical support to achieve the practical objectives identified within each text. It is expected that many institutions worldwide will also provide such support.

The project will continue at Thames Polytechnic to support these 'Open Learning Texts', to continually refresh and update the material and to extend its coverage.

Further information about nominated support centres, the material or open learning techniques may be obtained from the project office at Thames Polytechnic, ACOL, Wellington St., Woolwich, London, SE18 6PF.

How to Use an Open Learning Text

Open learning texts are designed as a convenient and flexible way of studying for people who, for a variety of reasons cannot use conventional education courses. You will learn from this text the principles of one subject in Analytical Chemistry, but only by putting this knowledge into practice, under professional supervision, will you gain a full understanding of the analytical techniques described.

To achieve the full benefit from an open learning text you need to plan your place and time of study.

- Find the most suitable place to study where you can work without disturbance.

- If you have a tutor supervising your study discuss with him, or her, the date by which you should have completed this text.

- Some people study perfectly well in irregular bursts, however most students find that setting aside a certain number of hours each day is the most satisfactory method. It is for you to decide which pattern of study suits you best.

- If you decide to study for several hours at once, take short breaks of five or ten minutes every half hour or so. You will find that this method maintains a higher overall level of concentration.

Before you begin a detailed reading of the text, familiarise yourself with the general layout of the material. Have a look at the course contents list at the front of the book and flip through the pages to get a general impression of the way the subject is dealt with. You will find that there is space on the pages to make comments alongside the text as you study—your own notes for highlighting points that you feel are particularly important. Indicate in the margin the points you would like to discuss further with a tutor or fellow student. When you come to revise, these personal study notes will be very useful.

Π When you find a paragraph in the text marked with a symbol such as is shown here, this is where you get involved. At this point you are directed to do things: draw graphs, answer questions, perform calculations, etc. Do make an attempt at these activities. If necessary cover the succeeding response with a piece of paper until you are ready to read on. This is an opportunity for you to learn by participating in the subject and although the text continues by discussing your response, there is no better way to learn than by working things out for yourself.

We have introduced self assessment questions (SAQ) at appropriate places in the text. These SAQs provide for you a way of finding out if you understand what you have just been studying. There is space on the page for your answer and for any comments you want to add after reading the author's response. You will find the author's response to each SAQ at the end of the text. Compare what you have written with the response provided and read the discussion and advice.

At intervals in the text you will find a Summary and List of Objectives. The Summary will emphasise the important points covered by the material you have just read and the Objectives will give you a checklist of tasks you should then be able to achieve.

You can revise the Unit, perhaps for a formal examination, by re-reading the Summary and the Objectives, and by working through some of the SAQs. This should quickly alert you to areas of the text that need further study.

At the end of the book you will find for reference lists of commonly used scientific symbols and values, units of measurement and also a periodic table.

Contents

Study Guide

This Unit is designed to introduce you to the basic theory and practice of atomic spectroscopy. The emphasis is on the practical and instrumental aspects and the use of mathematics and theoretical spectroscopy is kept to a minimum.

A description of the basic principles of atomic spectroscopy is given in Parts 1 and 2 of the Unit.

The basic instrumentation and practice of the (currently) most widely used technique in atomic spectroscopy – flame atomic spectroscopy – is described in Parts 3 and 4 of the Unit. In order to use atomic spectroscopy as an effective analytical tool it is important to understand the problems which can arise, and to know how to combat these problems by sample treatment or instrumental methods.

Thus the first four Parts of the Unit are designed to give you a basic grounding in the theory, instrumentation and practice of atomic spectroscopy.

The more specialised techniques of atomic absorption are described in Parts 5 and 6. Emission spectroscopy, which has been less widely used than absorption spectroscopy in the past, has become an increasingly more attractive method in recent years for reasons described in Part 7. Part 8 describes the relative merits of what may, to the newcomer, be a bewildering variety of methods, in order to give an overall perspective on the Unit.

Study Guide

Supporting Practical Work

The range of atomic spectroscopy instrumentation available will vary widely from one laboratory to another. The experiments which follow are designed to illustrate basic procedures and will not require particularly sophisticated instrumentation.

Aims

(*a*) To provide experience in the preparation and pre- treatment of samples in a form suitable for analysis.

(*b*) To provide experience of operating and otpimising appropriate instrumentation.

(*c*) To select a suitable working range for a given element, and construct and use calibration curves.

(*d*) To demonstrate the interferences which can arise, and methods of combating these interferences.

(*e*) To demonstrate the use of the method of standard additions.

Suggested Experiments

The experiments suggested here are designed to be carried out without the use of particularly specialised or expensive equipment.

(*a*) Optimisation of an atomic absorption spectrometer.

(*b*) The determination of potassium in tomato puree. (Using calibration curve and method of standard additions).

(*c*) The determination of magnesium in tap water. (With and without EDTA).

(*d*) The determination of iron in beer. (Using APDC/MIBK extraction).

(*e*) The determination of sulphate. (Precipitation with barium).

Experiments (*a*) to (*e*) can be carried out on flame atomic absorption spectrometers. Experiments (*b*) and (*e*) can also be carried out on flame photometers.

If a graphite furnace accessory is available, the following experiments are suggested:

(*f*) Optimisation of ashing and atomisation for cadmium and copper.

(*g*) Determination of arsenic using nickel chloride as a matrix modifier.

Note

In addition to selecting some of the above experiments, you may wish to use the computer simulation program provided by the CALM project.

Bibliography

1. ATOMIC SPECTROSCOPY TEXTBOOKS

(*a*) L Ebdon, *An Introduction to Atomic Absorption Spectroscopy*, Heyden, 1982.

(*b*) G F Kirkbright and M Sargent, *Atomic Absorption and Fluorescence Spectroscopy*, Academic Press 1974.

(*c*) W J Price, *Spectrochemical Analysis by Atomic Absorption*, Heyden, 1979.

(*d*) M Slavin, *Atomic Absorption Spectroscopy*, Wiley, 2nd Edn, 1978.

(*e*) M Thompson and J N Walsh, *A Handbook of Inductively Coupled Plasma Spectrometry*, Blackie 1983.

(*f*) B Welz, *Atomic Absorption Spectroscopy*, Verlag Chemie, 1976.

Notes

Reference (*a*) is readable, comprehensive and fairly up to date.

References (*c*) and (*d*) are practically-orientated.

Reference (*e*) is a specialist text.

2. 'STANDARD' ANALYTICAL TEXTBOOKS

Most comprehensive analytical chemistry texts will contain a chapter on atomic spectroscopy. Examples include:

(*a*) T Kuwana (Ed), *Physical Methods in Modern Chemical Analysis*, Vol 1, Academic Press, 1978.

(*b*) G Svehla (Ed), *Wilson and Wilson's Comprehensive Analytical Chemistry*, Elsevier, 1975 (several volumes).

(*c*) H H Willard, L L Merit, J A Dean and F A Settle, *Instrumental Methods of Analysis.* Van Nostrand, 1981.

3. JOURNALS

It is particularly true of atomic spectroscopy that text books soon date, because of the rapid developments in the subject. Some of the journals which contain reviews and original papers are given in the following list:

(*a*) *Analyst*

(*b*) *Analytica Chimica Acta*

(*c*) *Analytical Chemistry*

(*d*) *Annual Reports on Analytical Atomic Spectroscopy*

(*e*) *European Spectroscopy News*

(*f*) *International Laboratory*

(*g*) *Journal of Analytical Atomic Spectroscopy*

(*h*) *Spectrochimica Acta*

Journals (e) and (f) contain occasional review articles aimed at the general reader. The other journals contain original research papers, although some, such as (c) and (g) also publish general review articles. Journal (d) contains detailed and comprehensive reviews.

Acknowledgements

The author would like to thank most sincerely, Dr. Steve Haswell of Thames Polytechnic for his helpful advice in writing this text.

The frontispiece is two colour plates provided by Mr A Batho of the Analytical Instrument Division, Thermoelectron, Birchwood, Warrington, Cheshire WA3 7QZ.

Figure 2.3a is redrawn from *Analytical Proceedings*, 22, 63, 1985, with the permission of the Royal Society of Chemistry.

Figure 4.4b is redrawn from M L Parsons, B W Smith and G E Bentley, *Handbook of Flame Spectroscopy*, Plenum Press, 1977, with the permission of Plenum Publishing Corp.

Figures 5.2c and 5.2d are redrawn from an International Laboratory, Operator's Manual. Permission has been applied for.

Figures 7.2a and 7.2b are redrawn from the Corning catalogue: *A guide to Flame Photometer Analysis.* Permission has been applied for.

1. Atomic Spectroscopy – General Introduction

Overview

After a short historical introduction, some basic spectroscopic ideas essential to this Unit are revised. The main features of the different types of atomic spectroscopy are then reviewed. The scope of atomic spectroscopy is then discussed in terms of which elements are most conveniently determined.

1.1. INTRODUCTION

Atomic spectroscopy has its origins in the flame test in which many elements can be identified by the characteristic colours which their salts give to a flame.

Talbot, in 1826, saw that a wick impregnated with table salt (sodium chloride) burnt with the emission of an intense yellow light, and that the same colour was obtained with other sodium salts. Potassium salts, on the other hand, gave a different colour, a bluish tinge, to flames. The metals present in salts could in many cases be identified from the colours which they gave to flames.

Simple visual observation of flame colours was, however, rather limited. For example both lithium and strontium give red flames which are not easily distinguished. An improvement was to use a prism to disperse the light emission from the flame into its component wavelengths (Fig. 1.1a). Lithium and strontium are then easily identified since in the red region of the spectrum, lithium emits only one wavelength while strontium emits several different wavelengths.

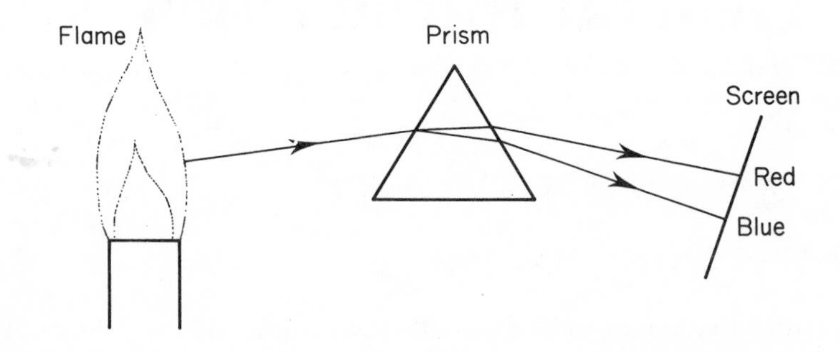

Fig. 1.1a. *Dispersion of light emitted from a flame*

With the development of the bunsen burner, giving a relatively hot and colourless flame, Bunsen and Kirchoff improved the technique. Several previously unknown elements were identified using the technique, for example caesium and rubidium (1861) and helium (1895).

As we shall see in this Unit atomic spectroscopy has made tremendous advances and is now a very widely used technique for both the identification and quantitative determination of many elements present in samples. There are many variations of the technique, but all share two main characteristics:

(*i*) *specificity* – individual elements in given sample can be reliably identified.

(*ii*) *sensitivity* – the amounts of an element that can be detected are very small. Levels of around 1 ppm (part per million) can be measured with straightforward procedures. Even smaller levels of 0.001 ppm or less can be measured with more sophisticated procedures.

1.2. SOME BASIC CONCEPTS IN SPECTROSCOPY

The Unit requires a good understanding of three basic concepts in spectroscopy, namely

(*i*) the wave nature of light;

(*ii*) the particle properties of light;

(*iii*) the absorption and emission of light.

This section covers material which may already be familiar to you. If you can answer the following three SAQs, you may skip the rest of this section and move on to Section 1.3. If you have some difficulty in answering these SAQ's you will need to study this section and then attempt the questions again.

SAQ 1.2a

The characteristic yellow light emission of sodium, seen in flames containing sodium salts, or from sodium street lamps, occurs at a wavelength of 589 nm.

For photons of this wavelength, calculate

(*i*) the frequency;

(*ii*) the wavenumber;

(*iii*) the energy.

(i) $\nu = c/\lambda = \dfrac{3 \times 10^{8}\ m\ s^{-1}}{5.89 \times 10^{-7}\ m}$

$= 5.09 \times 10^{14}\ s^{-1}\ (Hz)$

(ii) $\bar{\nu} = \dfrac{1}{5.89 \times 10^{-7}\ m} = 1.70 \times 10^{6}\ m^{-1}$

$\bar{\nu} = 1.698 \times 10^{4}\ cm^{-1}$

(iii) $E = \nu h$

$E = (5.09 \times 10^{14}\ s^{-1})(6.626 \times 10^{-34}\ J\ s)$

$= 3.37 \times 10^{-19}\ J$

$E\ (6.022 \times 10^{23}\ mol^{-1}) = 203\ kJ/mol$

SAQ 1.2b

Arrange the following regions of the electromagnetic spectrum in order of (*a*) increasing wavelength, and (*b*) increasing energy:

(*i*) ultra-violet;

(*ii*) infra-red;

(*iii*) visible.

SAQ 1.2c

Fig. 1.2a, (*i*), (*ii*) and (*iii*) depict the three energy level diagrams for the processes of absorption of radiation, emission of radiation and fluorescence. Identify each process.

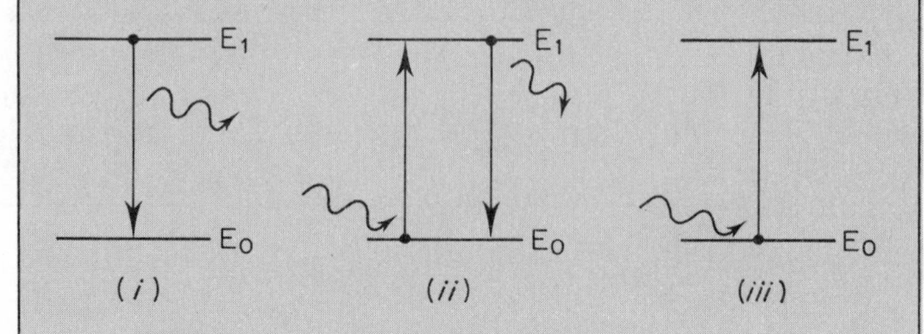

Fig. 1.2a. *Energy level diagrams for absorption, emission and fluorescence*

i) emission

ii) fluorescence

iii) absorption

1.2.1. The wave nature of light

The observations that light can give rise to interference patterns, diffraction etc., show that light must behave like a wave. According to the electromagnetic theory of radiation, developed by Maxwell in the 19th century, light consists of photons which can be thought of as oscillating electric fields travelling very rapidly through space. Associated with the electric field (E) is a magnetic field (B). The fields vibrate at right angles to each other and also at right angles to the direction in which the photon is moving, as shown in Fig. 1.2b.

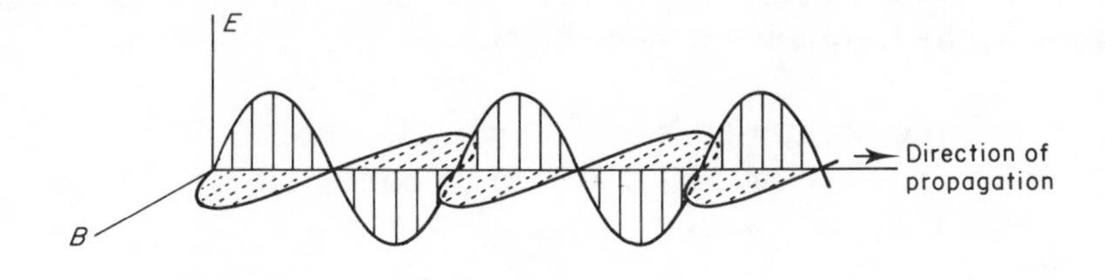

Fig. 1.2b. *Electromagnetic radiation*

] For the purposes of this Unit the magnetic field can be ignored as it is mainly of importance in inducing magnetic transitions, such as in nmr spectroscopy.

As the photon oscillates like a sine wave, the photon can be described in terms of the characteristic properties of sine waves. These include the repeat distance of the wave, called the ***wavelength***, λ, (pronounced 'lambda') which is the distance between successive peaks or troughs as shown in Fig. 1.2c.

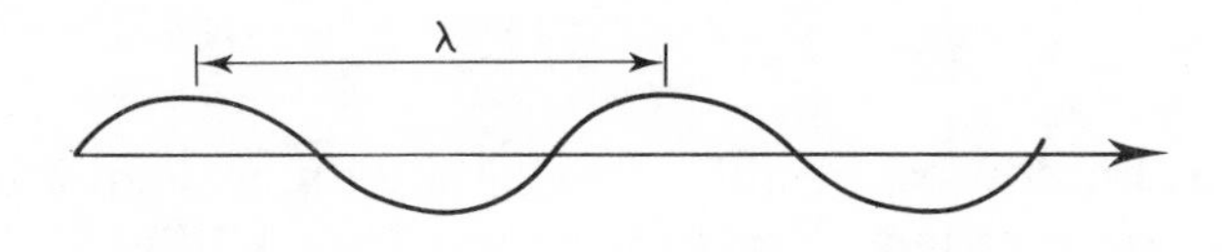

Fig. 1.2c. *The wavelength of a photon*

The number of times the wave oscillates in one second is called the *frequency*, ν, (pronounced 'nu'), of the photon. The frequency is the number of wavelengths in the distance travelled by the wave in one second, or the velocity, c, divided by the wavelength, in units of s^{-1} or Hertz (Hz)

$$\nu = c/\lambda \qquad (1.1)$$

or

$$c = \nu\lambda \qquad (1.2)$$

The velocity of light is a constant, 3×10^8 ms^{-1}, so if the wavelength is known, the frequency may be calculated.

Π For the sodium emission referred to in SAQ 1.2a, λ = 589 nm or 5.89×10^{-7} m (1 nanometre = 10^{-9} m), so you should now try to calculate ν.

The answer is obtained simply by substituting for velocity and wavelength in Eq. 1.1:

$$\nu = 3.0 \times 10^8/5.89 \times 10^{-7}$$

so

$$\nu = 5.093 \times 10^{14} \text{ Hz}$$

Note how high the frequency is (509 million million oscillations in a second), and how small the wavelength is (only about a few hundred times larger than an atom). Another useful way to characterise a wave is the number of waves in one metre, the ***wavenumber***, given the symbol $\bar{\nu}$ (pronounced 'nu-bar'). The wavenumber is simply equal to the reciprocal of the wavelength – if the wavelength were 0.1m there would be 10 waves in one metre – or:

$$\bar{\nu} = 1/\lambda \qquad (1.3)$$

Π You should now be able to calculate the wavenumber for the photons emitted by sodium atoms (SAQ 1.2a).

If the wavelength is 589 nm then using Eq. 1.3:

$$\bar{\nu} = 1/5.89 \times 10^{-7}$$

or $$\bar{\nu} = 1.698 \times 10^{6} \text{ m}^{-1}$$

Although the SI unit for the wavenumber is m^{-1}, the numerical values are rather large and cumbersome, and it is more usual to use the units of cm^{-1}.

Π Can you convert the above value of $\bar{\nu}$ in m^{-1} to cm^{-1}?

Some people find this conversion confusing. It may help to think of the SI unit as 'per metre'. Since the metre is a hundred times longer than the centimetre, there will be a hundred times as many waves 'per metre' compared with 'per centimetre'.

ie $$1 \text{ cm}^{-1} = 100 \text{ m}^{-1}$$

or $$1 \text{ m}^{-1} = 0.01 \text{ cm}^{-1}$$

The wavenumber for the sodium emission is then $1.698 \times 10^{6}/100$ or 1.698×10^{4} cm^{-1}.

1.2.2. The particle nature of light

Around the end of the 19th century, many limitations of the wave theory of light had become apparent. The theory was unable to account for such diverse phenomena as black body radiation and the photoelectric effect. Planck was able to explain these phenomena by suggesting that a photon has a particle nature, and that the energy of a photon is directly proportional to the frequency of the photon:

$$E \propto \nu \qquad 1.4$$

or

$$E = h\nu \qquad 1.5$$

where h, the Planck constant, is 6.626×10^{-34} J s.

Π Since you have already calculated the frequency for the photons emitted by sodium (SAQ 1.2a), you should now try to calculate the energy.

Your answer should be 3.375×10^{-19} J (that is $6.626 \times 10^{34} \times 5.093 \times 10^{14}$).

This is a rather small quantity, as it is the energy of only one photon. Chemists usually prefer to work with more familiar units - J mol^{-1} - obtained by multiplying the photon energy by the Avogadro constant N_A. Therefore

$$E = 3.375 \times 10^{-19} \times 6.022 \times 10^{23}$$

or

$$E = 203{,}000 \text{ J mol}^{-1} \text{ (203kJ mol}^{-1}\text{)}$$

We have now covered the material necessary to completely answer SAQ 1.2a.

You may also be able to answer SAQ 1.2b with the help of the following information:

The visible spectrum ranges from a wavelength of 400 nm at the blue end to 700 nm at the red end. The ultra-violet region is of lower wavelength than the blue of the visible region and the infrared region is of longer wavelength than red light. This is illustrated in Fig. 1.2d.

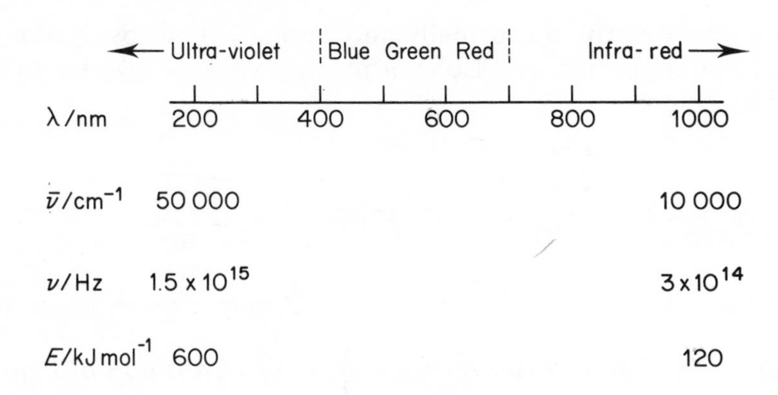

Fig. 1.2d. *The electromagnetic spectrum near the visible region*

1.2.3. Absorption and emission of light

Since a photon behaves rather like an alternating electric field, it can interact with the negatively charged electrons in an atom. Under certain conditions this interaction can lead to a photon being *absorbed* by an atom. The energy levels in an atom are quantised, that is they can only have certain well-defined energies. An important consequence of this is that the photon energy, hν, must be exactly equal to the energy separation between a filled energy level (E_0) and an unoccupied energy level (E_1), as shown in Fig. 1.2e.

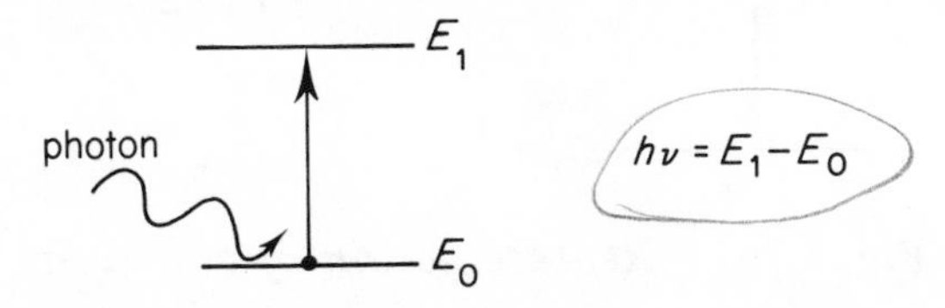

Fig. 1.2e. *Representation of an absorption process*

The wave in Fig. 1.2e represents a photon colliding with and being absorbed by an atom in its ground or stable state E_0. The straight arrow represents the simultaneous (on a time scale of 10^{-15}s) excitation of the atom from E_0 to the excited state E_1. This process involves an electron being promoted from a filled atomic orbital to a more energetic orbital (normally unoccupied). The absorption process can be described by a Bohr orbit diagram as shown for lithium in Fig. 1.2f.

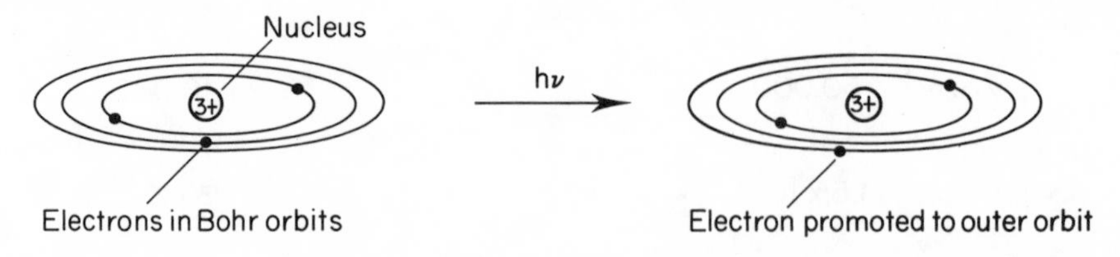

Fig. 1.2f. *Bohr orbit representation of absorption of light by a lithium atom*

(Note that this is a rather over-simplified representation of absorption, since transitions between different orbitals within the same electronic shell can be induced, as well as transitions between different electronic shells. An example of this is the promotion of an electron in a 3s orbital in sodium atom to the 3p orbital by light of wavelength 589 nm.)

The process of *emission* is the exact reverse of absorption. An atom in an excited energy level can revert to the ground energy level by emitting the extra energy as a photon, as shown in Fig. 1.2g.

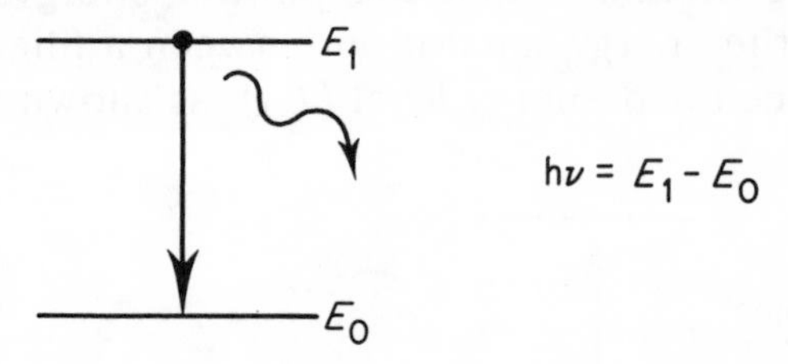

Fig. 1.2g. *Representation of emission*

Again the photon energy $h\nu$ is equal to the energy gap ($E_1 - E_0$). In the sodium emission at 589 nm, electrons in 3p orbitals are returning to 3s orbitals. The transition may be represented

$$\text{Na [Ne](3p)}^1 \rightarrow \text{Na [Ne](3s)}^1$$

where [Ne] indicates the filled electron sub-shells in the atom.

Although less common, the third important process involves a combination of the above two processes with an absorbed photon being almost immediately re-emitted (Fig. 1.2h). This is called *fluorescence*.

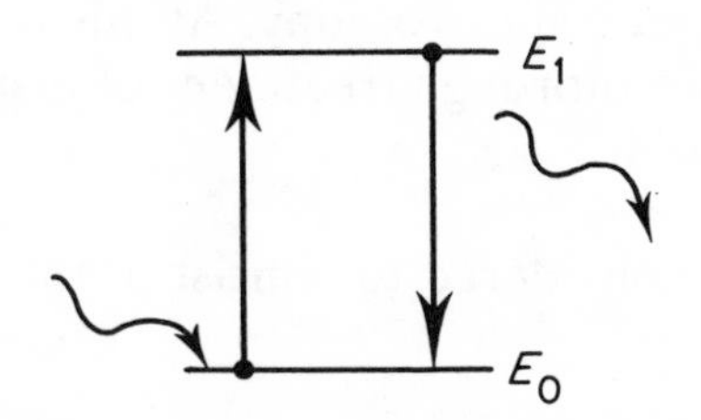

Fig. 1.2h. *Representation of fluorescence*

Fluorescence may also be represented as follows:

$$\text{Atom} + h\nu_1 \rightarrow (\text{Atom})^* \rightarrow \text{Atom} + h\nu_2$$

where the asterisk indicates that the atom is electronically excited, $h\nu_1$ represents the photon absorbed and $h\nu_2$ the photon emitted.

In the special case illustrated in Fig. 1.2h, where the absorption and emission are between the same two energy levels the process is called *resonance fluorescence* and $h\nu_1 = h\nu_2$. Normally $h\nu_1$ is greater than $h\nu_2$.

The three types of photon-atom interactions covered in this Section form the basis of the three main branches of atomic spectroscopy, namely

atomic absorption spectroscopy (AAS)

atomic emission spectroscopy (AES), and

atomic fluorescence spectroscopy (AFS)

You should now be able to answer SAQ 1.2c.

1.3. THE MAIN COMPONENTS OF ATOMIC ABSORPTION SPECTROMETERS

In this Section we look at the essential spectrometer components, and then look at how they can be combined for the three different forms of atomic absorption spectroscopy. At this stage our treatment will be brief, but a more thorough treatment of instrumentation will be given in Parts 3 and 4.

Spectrometers can be considered to consist of three parts:

(*a*) the light source;

(*b*) the atom cell;

(*c*) the light detection system (monochromator, detector and read-out units.)

We shall see that the instrumental requirements of the techniques of AAS, AES and AFS are rather different.

1.3.1. The light source

As in many forms of spectroscopy, there are several types of light source used. The source needed will depend on the type of atomic spectroscopy used.

Π Which of the three techniques described in Section 1.2.3 – AAS, AES or AFS – does not require a light source?

The answer is atomic emission spectroscopy (AES), since the atoms form the light source themselves. We only need to analyse the light emitted from the atoms of interest in the atom cell. An external source is needed for AAS where the absorbing atoms, which must be in the ground electronic state, in the atom cell reduce the intensity of the radiation from the source. In AFS the light emitted from atoms in the atom cell is measured, as in AES, but a light source is also needed since the atoms need to be excited to high atomic energy levels by light absorption.

The most widely used light sources in AAS and AFS are lamps which contain the element to be analysed. These lamps emit spectral lines of exactly the same energy as the absorbing atoms in the atom cell. These are usually *hollow cathode lamps* in the case of AAS and *electrodeless discharge lamps* in the case of AFS.

1.3.2. The atom cell

The *atom cell* is the name given to the part of the spectrometer in which the atoms are formed in the gaseous state. It is therefore the most important part of the spectrometer. The materials analysed in atomic spectroscopy are generally in the form of solids or solutions, in which the element of interest is bonded, either covalently or ionically to other elements. To obtain the atomic form of the element, we have to have a way of breaking the bonds.

Π Select the most likely condition to favour atomisation from the list below:

(*i*) low temperature;

(*ii*) low pressure;

(*iii*) high temperature;

(*iv*) high pressure.

Chemical bonds, whether ionic or covalent, are formed from the elements with the evolution of heat ie exothermically. In order to break the bonds in a compound heat must be put in to the compound as the process is endothermic. The correct answer to this question is then (*iii*) – high temperature. Answers (*i*) and (*iv*) are definitely wrong, since both conditions will tend to prevent dissociation of the compound. Answer (*ii*) is not incorrect, since low pressures favour dissociation of molecules (le Chatelier's principle), but this effect is not in itself of importance except at high temperatures.

The temperatures needed for atomisation are in fact very high, typically 2000 K to 5000 K. *Flames* can be used to obtain temperatures of 2000 K to 3500 K. Even higher temperatures of 5000 K to 10000 K can be formed in *plasmas* which are formed by electrical discharges.

1.3.3. The detector

In most cases more than one wavelength of radiation will be emitted from the light source and the atom cell. Before analysis of the emitted or absorbed light, some form of monochromator is necessary. The spectral line of interest is rarely the only light leaving the atom cell. Both atomic and molecular emission from other atoms and molecules are usually present. These can come from the lamp or the flame.

The detector normally used is a photomultiplier since this is very sensitive and has a fairly uniform response to light over the whole visible-uv region.

The photomultiplier amplifies the signal but a further amplifier is normally also used, and some form of readout system is required. This may be a pen recorder, a meter, or a computer + printer.

1.3.4. Comparison of AAS, AES and AFS experiments

The principal features of the three techniques are described in Fig. 1.3. All techniques require an atom cell, a monochromator, detector and readout. The differences lie in the light source and the geometry.

Π Can you label the diagrams (*a*), (*b*) and (*c*) in Fig. 1.3 as corresponding to the techniques of AAS, AES and AFS?

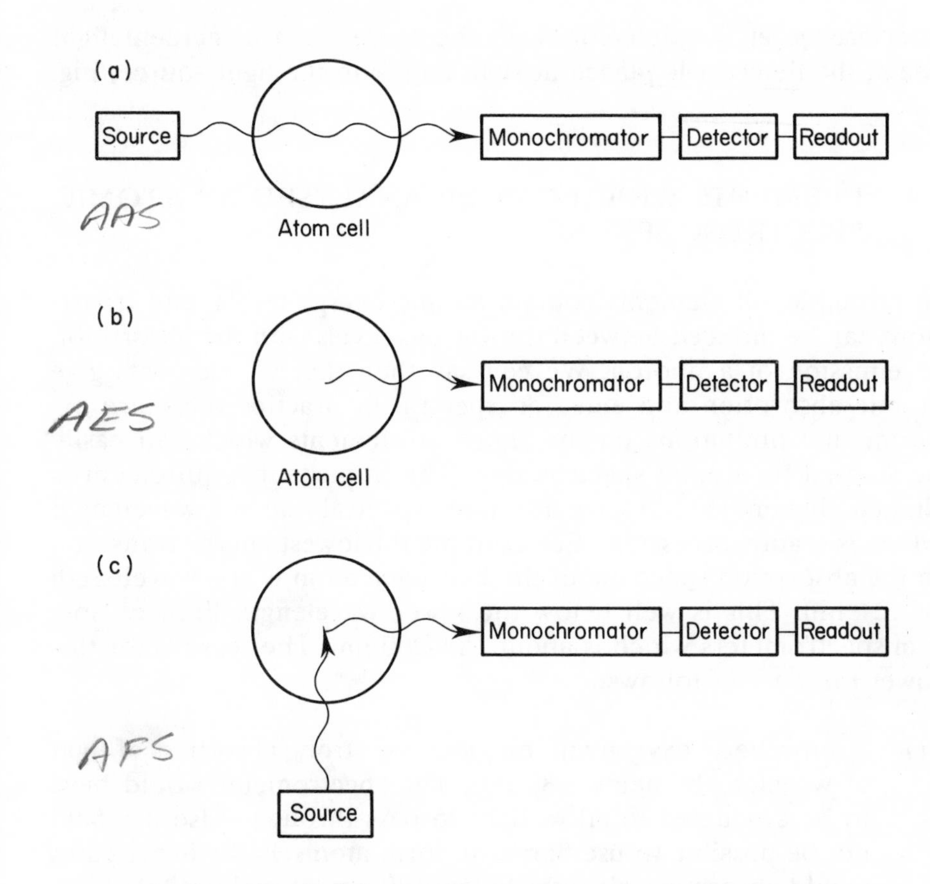

Fig. 1.3. *Main features of atomic spectroscopy techniques*

We have already seen in Section 1.3.2 that emission spectroscopy does not require a light source. Therefore Fig. 1.3b corresponds to AES.

The difference between the other techniques is in the geometry. For absorption the light must pass straight through the atom cell to the detector, so Fig. 1.3a corresponds to AAS.

In fluorescence experiments the fluorescence occurs in all directions, so in principle the fluorescence could be detected at any angle. In practice the fluorescence signal is weak because very little of the emitted light can be collected in any one direction. To avoid interference at the detector from the more intense incident light beam, the detector is placed at right angles to the light source, Fig. 1.3c.

1.4. ELEMENTS WHICH CAN BE ANALYSED BY ATOMIC SPECTROSCOPY

In principle, all elements contain atomic energy levels, and transitions can be induced between the various levels with the absorption or emission of a photon. We can say then that all elements give atomic absorption and emission spectra. In practice there are experimental limitations on the range of elements which can easily be studied by atomic spectroscopy. The important requirement is that an element should have a suitable spectral line at a wavelength which is readily accessible. For example the lowest energy transition in the absorption spectrum of the hydrogen atom is at a wavelength of 121 nm. This is well below the lower wavelength limit of normal spectrometers which is about 180–200 nm. The reasons for this lower limit are as follows.

(*i*) Atmospheric oxygen and nitrogen are strong absorbers of light of wavelengths below 180 nm. The spectrometer would have to be evacuated to allow light to pass through. Also it would not be possible to use flames to form atoms as the flame gases would absorb wavelengths below 180 nm strongly. While 'vacuum ultra-violet' spectrometers are in use, they are highly specialised pieces of equipment.

(*ii*) Glass, quartz and many plastic materials which might be used for optical components strongly absorb light below 180 nm. Materials such as alkali halide crystals, which are not so cheap or easy to handle, would have to be used.

(*iii*) Light sources below 200 nm usually have very weak output, a consequence of the very high energies of photons having these wavelengths.

The long wavelength limit is determined by the availability of sufficiently sensitive detectors. The most widely used type of detector is the photomultiplier which can be used at wavelengths up to about 1000 nm, but beyond this wavelength the photons do not have enough energy to cause ionisation in the detector (see Part 3).

The accessible wavelength range then is 180–1000 nm. Most, but not all, elements give atomic absorption lines in this part of the spectrum. Taking the first row elements as an example, lithium at one end of the row gives accessible absorption lines, fluorine at the other end of the row does not. This is because there is a trend across the row for the energy level separations to increase. This is a consequence of the increasing nuclear charge leading to stronger interactions between the electrons and the nucleus. The increasing strength of these electrostatic forces is also shown by the very much higher ionisation potential of fluorine (1681 kJ mol^{-1}) compared to lithium (520.3 kJ mol^{-1}). The difference in the relative energies of the two absorption transitions is illustrated in Fig. 1.4a:

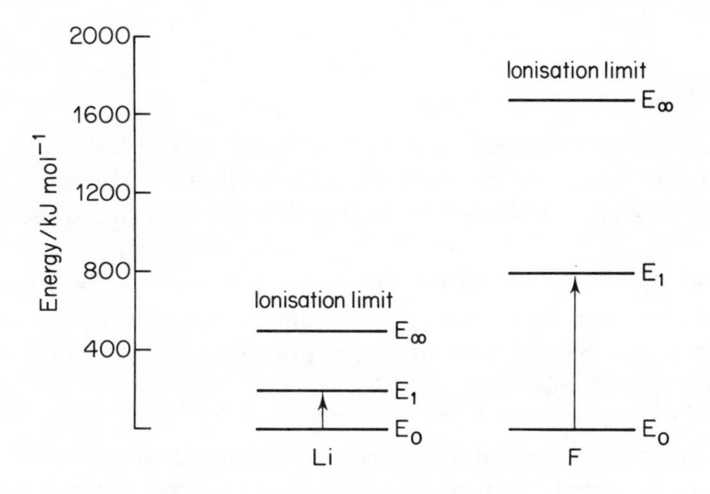

Fig. 1.4a. *Absorption transitions for Li and F*

We can see that the energy of the absorption transition is much higher for fluorine (below 200 nm) than for lithium (above 200 nm).

The trend across the periodic table becomes less marked for heavier elements, as the valence electrons are much further away from the nucleus in the larger atoms. The heavier elements at the right hand side of the periodic table are more likely to give absorption lines at accessible wavelengths than the lighter atoms. Iodine, for example, has a useful absorption line at 206 nm. A useful rule of thumb is that metallic elements all give accessible absorption lines. (The fact that they are metals means that the valence electrons are not held strongly.) Alternatively, we can remember that it is the elements at the top right hand corner of the periodic table which are not readily analysed by atomic absorption spectroscopy.

Π Which of the following elements would you expect to give accessible atomic absorption lines?

helium

potassium

calcium

oxygen

Both potassium and calcium are, like lithium, at the left hand side of the periodic table and have easily accessible excited states. They are therefore readily analysed by atomic absorption spectroscopy.

Oxygen and helium are both on the right hand side of the periodic table and of low atomic number. The atomic absorption lines at 131 nm (oxygen) and 58 nm (helium) are not readily accessible using commercial spectrometers.

So far we have only considered absorption spectroscopy. The case for emission is slightly different in that observed transitions do not have to go back to the ground state, but can go to another excited state. The excited states are usually closer to each other than to the ground state, so transitions between excited states can be at wavelengths which are accessible. For example, in the case of the hydrogen atom the transition in which an electron moves from the 3p orbital to the 2s orbital results in light emission at 656 nm. The only problem with observing this type of emission is that we have to put a great deal of energy into the atom in order to generate the highly excited atomic states. The amount of energy available in flames is normally too low to populate highly excited states. (These states can be populated however in some energetic electrical discharges, known as plasmas.)

To summarise then, most of the elements in the periodic table can be studied, and it is the metallic elements which are most readily analysed by both atomic absorption and atomic emission spectroscopy. Since nearly all the heavier elements are metallic, well over half the elements in the periodic table can be determined by routine methods. The non-metallic elements can actually be determined by atomic emission spectroscopy but very high temperatures are needed to form excited atoms of these elements.

SUMMARY AND OBJECTIVES

Summary

Atomic spectroscopy can be used in three different forms – absorption, emission and fluorescence. All are highly specific to individual elements, and are very sensitive techniques. For transitions to be observed the emitted or absorbed photon must be of exactly the same energy as the energy gap between the ground and excited states. High temperatures are needed to generate atoms. In principle, all elements in the periodic table can be analysed by atomic spectroscopy, but in practice the technique lends itself more readily to the metallic elements.

Objectives

You should now be able to:

- appreciate that for absorption or emission the photon energy must exactly equal the energy of the atomic transition;

- relate wavelength, wavenumber, energy and frequency of photons to each other;
- explain the differences between absorption, emission and fluorescence;
- describe the main features of the three types of atomic spectroscopy;
- decide which of the elements in the periodic table are likely to be most readily analysed by atomic spectroscopy.

λ unit m

$\bar{\nu} = \frac{1}{\lambda}$

$E = h\nu$

$\nu = c/\lambda$

2. The Theory of Atomic Spectroscopy

Overview

It is possible to use atomic spectroscopy without knowing the main principles, for carrying out a routine analysis. A 'cookbook' recipe which has already been developed can be followed. Even with routine analyses, however, unforseen problems can arise. To tackle these problems, some understanding of the theory of the spectral transitions being studied, and of the instrumentation being used is needed. In this Section we study the most relevant aspects of the theory:

(*a*) the origins of spectral transitions;

(*b*) the populations of energy levels;

(*c*) the factors influencing the spectral line widths.

The importance of these aspects of the theory will be apparent in the later Parts where individual techniques are studied in more detail. The necessary theory of instrumentation will be described in Parts 3 and 4. The level of mathematics required will be minimal. The results of quantum theory will be discussed in a qualitative manner.

2.1. THE ORIGINS OF SPECTRAL TRANSITIONS

The spectral transitions involved in AAS are electronic transitions, in which electrons move from one atomic orbital to another with the simultaneous absorption (or emission) of light. The origins of these transitions will be described for sodium, which has only one valence electron. The electronic transitions for atoms with more than one valence electron are more complex, as we shall illustrate for magnesium.

2.1.1. The sodium atom

The visible emission spectrum of atomic sodium includes an intense line near 590 nm and another line near 569 nm. On closer inspection at high resolution the intense line is found to be a doublet, that is it consists of two lines very close together at 589.0 nm and 589.6 nm.

A simplified energy level diagram for the sodium atom is shown in Fig. 2.1a. The s, p, d and f orbitals within a shell (with the same principal quantum number) do not have the same energy. This is due to the electron-electron interactions which are present in all atoms containing more than one electron. For the n = 3 shell:

$$E\,(3d) > E\,(3p) > E\,(3s)$$

The emission lines mentioned above are due to the 4d $\rightarrow$ 3p and 3p $\rightarrow$ 3s transitions as shown on the diagram.

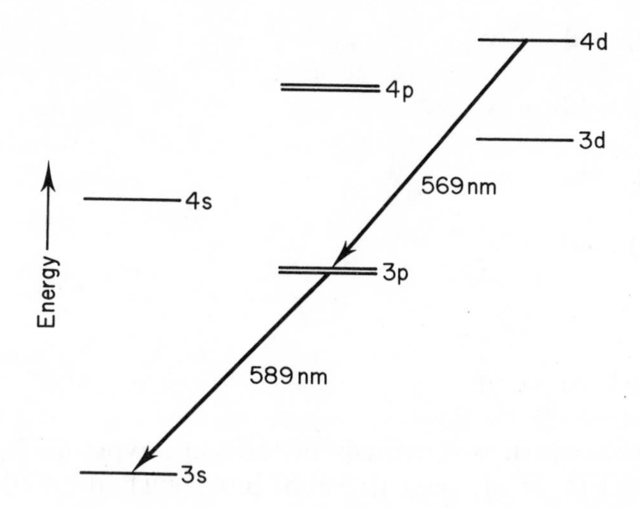

Fig. 2.1a. *Energy levels for sodium (simplified)*

Not all possible transitions are allowed. This is because there is a selection rule which allows only transitions between levels in which the orbital angular momentum quantum number, l, changes by one unit:

$$\Delta l = +1 \text{ or } -1$$

The values of l for the different orbitals are:

		s	p	d	f
l	=	0	1	2	3

The transitions shown on the diagram are both allowed since in each case l changes by one unit.

Π Select which of these other possible transitions between the levels shown in Fig. 2.1a are allowed.

(*i*) 4s → 3s

(*ii*) 4p → 3s

(*iii*) 3d → 3p

(*iv*) 3d → 3s

(*v*) 4d → 4s

(*vi*) 4d → 3s

(*vii*) 4p → 3p

(*viii*) 4d → 3d

Transitions between two orbitals of the same type are forbidden, therefore (*i*) (s → s), (*vii*) (p → p) and (*viii*) (d → d) are not allowed.

Transitions in which l changes by 2 units are also forbidden, therefore (*iv*), (*v*) and (*vi*) (all d → s) are not allowed.

The two remaining possibilities are both allowed. (*ii*) (4p → 3s) is observed in the ultraviolet at 330 nm. (*iii*) (3d → 3p) is observed in the infrared at 819 nm.

Although the process of emission is the reverse of absorption, there is an important difference between the absorption and emission spectra as the next question should show.

Π Which transitions between the energy levels shown in Fig. 2.1a would you expect to see in the absorption spectrum of sodium atoms?

Two absorption transitions are observed, from the 3s to the 3p orbital, and from the 3s to the 4p orbital. As for emission the selection rule $\Delta l = 1$, limits the number of possible transitions. There is a further restriction, namely that absorption transitions are only observed from the ground state (from the 3s orbital). We shall see in Section 2.2 that the concentration of excited atoms is always many orders of magnitude less than the concentration of ground state atoms, so that absorption of light by excited atoms will be negligible.

There are four observed emission transitions and only two allowed absorption transitions from the list given above. These observations are summarised in Fig. 2.1b.

Transition	Wavelength	Absorption	Emission
3s↔3p	589 nm	Yes	Yes
3s↔4p	330 nm	Yes	Yes
3p↔3d	819 nm	No	Yes
3p↔4d	569 nm	No	Yes

Fig. 2.1b. *Observed transitions for sodium*

Emission spectra are always more complex than absorption spectra, with an important consequence. As we shall see in more detail in Section 3, the light sources used in atomic absorption spectroscopy are basically atomic emission sources of the analyte element. Not all the lines emitted by an element can be re-absorbed by that element. Only those emission lines which result in the excited atom returning to the ground electronic state can be re-absorbed. These lines are called resonance lines.

Π Fig. 2.1c describes a partial energy level diagram for lead atoms. The ground electronic energy level is E_0. Four of the observed emission lines from energy level E_4 are shown. Which of these emission lines are resonance lines?

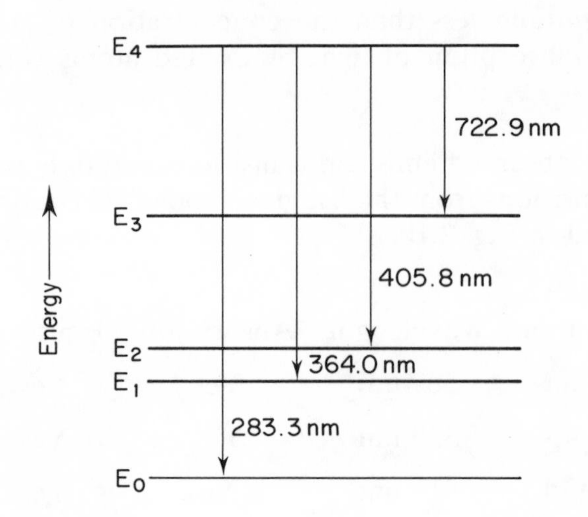

Fig. 2.1c. *Partial energy level diagram for Pb*

There is one resonance line – 283.3 nm. This is the only transition in which the atom returns to the ground state.

The doublet fine structure observed for the emission line of Na at 589 nm is rather more difficult to explain. It arises because not all the electrons in p orbitals have exactly the same energy, due to the interaction of the electron spin momentum with the orbital angular momentum. These magnetic interactions between the electron and the orbital result in the formation of different states of slightly different energies. Thus for Na atoms the 3p energy level is split into two states. The higher energy 3p state gives rise to the 589.6 nm spectral line. The other state has a slightly lower energy and give rise to the 589.0 nm spectral line.

The electron – orbital interaction is called Russell-Saunders coupling and becomes more important for heavier atoms. For alkali metals further down the periodic table, doublets are observed in which the separation is very much larger. (In the above example for sodium the energy difference between the two states is 200 J mol^{-1}. For caesium the energy difference is about 4000 J mol^{-1}!)

2.1.2. The magnesium atom

The alkaline earth metals, contain two electrons in the outer shell. Because of the extra valence electron the energy levels are more complex than for the alkali metals. Fig. 2.1d shows a partial energy level diagram for magnesium.

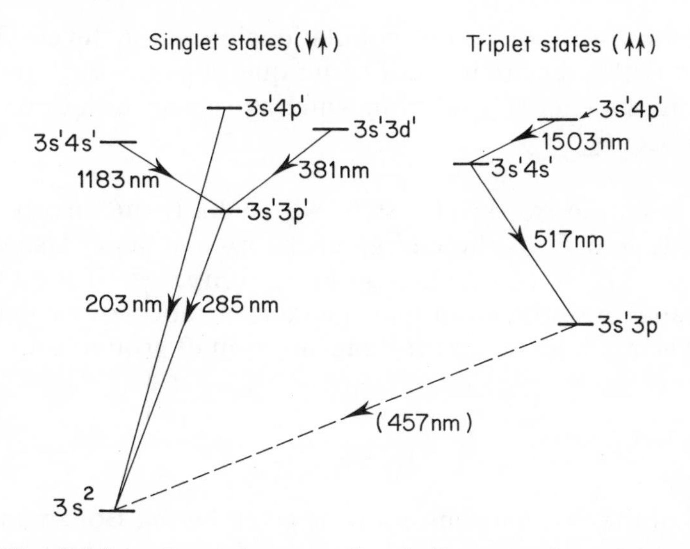

Fig. 2.1d. *Partial energy level diagram for magnesium*

Two sets of energy levels are obtained. In one set the electrons are paired with anti-parallel spins, and are known as singlet states. In the other set the electrons have parallel spins and are known as triplet states. Absorption or emission transitions are more likely between states of the same type, as the time needed for an electron to change its spin is usually much longer (about 10^{-9}s) than the time for a photon to be absorbed or emitted.

2.1.3 Other atoms

The complexity of the electronic structure of atoms increases with the number of valence electrons. The complexity is greatest for a half-filled shell, particularly for transition metals where there are a large number of electrons in the partially filled shell, resulting in a variety of electronic states. The emission spectrum of iron atoms, for example, contains thousands of lines in the uv-visible region.

2.2 THE INTENSITIES OF EMISSION AND ABSORPTION SPECTRAL LINES

In this Section we look at the populations of energy levels. This is important as the sensitivity of a technique depends on how many atoms are in the energy level from which the absorption or emission transition occurs.

Consider a simple two-level system where E_0 is the energy of the ground state and E_1 is the energy of the excited state. Usually the energy gap, $\Delta E = (E_1 - E_0)$, is so large compared to the thermal energy available to the atom that the concentration of excited state atoms N_1 is much less than the concentration of ground state atoms N_0:

$$N_1 << N_0$$

The ratio of the two concentrations is given by the Boltzmann Distribution Law;

$$N_1/N_0 = (g_1/g_0)\exp(-\Delta E/RT) \qquad 2.1$$

R is the molar gas constant 8.314 J K^{-1} mol^{-1}, and T is the absolute temperature.

The exponential term on the right hand side of the equation is the most important one, as it may vary by many orders of magnitude. The terms g_1 and g_0 are the degeneracies, (or the number of energy levels having the same energy), of the excited and ground states. g_1 and g_0 are usually small integers. In the case of sodium, as we discussed in the previous Section, there are two approximately degenerate energy levels in the excited state (when the electron is in a 3p orbital), so that $g_1 = 2$. There is only one ground state level, when the electron is in 3s orbital, so that $g_0 = 1$. For sodium then the equation simplifies to:

$$N_1/N_0 = 2 \exp(-\Delta E/RT)$$

At flame temperatures near 2000 K the ratio of concentrations of the excited and ground states can be calculated for the transition at 589 nm. By combining Eq. 1.1 and Eq. 1.5 we can see that the energy of a photon is inversely related to the wavelength

$$E = hc/\lambda$$

or, for 1 mole of photons:

$$E = hcN_A/\lambda \qquad 2.2$$

The numerator contains only fundamental constants and can be evaluated as

6.626×10^{-34}/J s $\times$ 2.998×10^{8}/m s^{-1} $\times$ 6.022×10^{23} mol^{-1}, or 0.1196 J m mol^{-1}.

Therefore, expressing λ in m:

$$E = 0.1196\ /589 \times 10^{-9}$$

$$= 203,060 \text{ J mol}^{-1}$$

and $$RT = 8.314 \times 2000$$

$$= 16,628 \text{ J mol}^{-1}$$

therefore $$N_1/N_0 = 2\exp(-203,060/16,628)$$

$$= 9.94 \times 10^{-6}$$

So even at flame temperatures only about one atom in 100 000 is in the excited state.

Π Temperatures in plasmas are much higher than in flames. Calculate the ratio N_1/N_0 for sodium at a plasma temperature of 5000 K.

The energy gap is the same as above, and RT is now 8.314 × 5000 = 41,570 J mol^{-1}, so:

$$N_1/N_0 = 2\exp(-203{,}060/41{,}570)$$

$$= 1.51 \times 10^{-2}$$

In this case, about 1.5% of the atoms are now in the excited state.

The intensity of absorption, I_a, is directly related to the concentration of atoms in the ground state, N_0.

The intensity of emission, I_e, is directly proportional to the number of atoms in the excited, emitting state, N_1.

The Boltzmann distribution law can be used to predict the effect of temperature and energy gap on the intensities of emission and absorption processes.

Π Which of the following statements are correct?

(*i*) Emission intensity increases with temperature.

(*ii*) Absorption intensity increases with temperature.

(*iii*) Absorption intensity is independent of the energy gap E.

(*iv*) Emission intensity increases with the energy gap.

(*v*) Emission intensity is independent of the wavelength of the photon emitted.

Statement (*i*) is correct.

Statement (*ii*) is incorrect.

Statement (*iii*) is correct

Statement (*iv*) is incorrect.

Statement (*v*) is incorrect.

The important relationships here are that the emission and absorption intensities are proportional to N_1 and N_0 respectively. That is:

$$I_e \propto N_1 \text{ and}$$

$$I_a \propto N_0$$

From the Boltzmann equation we can see that N_1 increases as temperature increases. Therefore I_e increases as temperature increases, and statement (*i*) is correct.

At temperatures below 10 000 K, $N_1 << N$, the total number of atoms, so that:

$$N_0 \approx N \approx \text{constant}$$

Therefore I_a is independent of temperature and statement (*ii*) is incorrect. (Changes in such factors as atomisation efficiency with temperature are ignored in this question.)

The effect of energy gap is opposite to the effect of temperature, the larger the energy gap at a particular temperature then the smaller is N_1. Hence I_e decreases as the energy gap increases and statement (*iv*) is incorrect.

Again N_0, and therefore I_a, will be independent of energy gap (as long as the energy gap is $>> RT$), and statement (*iii*) is correct.

The emission intensity decreases with increasing energy gap, and therefore increases with the wavelength of the emitted photon ($\Delta E \; \alpha \; 1/\lambda$), so that statement ($v$) is incorrect.

Two general rules emerge:

(a) The longer the wavelength of the photon emitted or absorbed, (the smaller the energy gap) the more likely emission spectroscopy is to be more sensitive, and therefore the technique of choice.

As a guide, if the analysis wavelength is in the infra-red, atomic emission spectroscopy is the most sensitive technique; if the wavelength is in the ultra-violet, atomic absorption spectroscopy is the most sensitive technique. For wavelengths in the visible region, emission is usually a little more sensitive than absorption, but either technique is probably satisfactory.

(b) The higher the temperature, the more sensitive the emission technique will be. (Absorption is less temperature dependent but higher temperatures may be favoured for more effective atomisation.)

Even when $N_1 << N_0$, emission spectroscopy is potentially a more sensitive technique than absorption spectroscopy for instrumental reasons. Basically, there is little background signal at the detector in emission experiments, so that very small currents can be measured. This is because when no analyte atoms are present in the atom cell no photons reach the detector. In absorption experiments, on the other hand, the background current is large as all the light from the source reaches the detector when no analyte atoms are present (Fig. 2.2a). The concentration is related to the difference signal with and without sample, which is small and liable to inaccuracies when the concentrations are low. This is essentially the same problem as arises in weighing by difference, where the actual weighings may be very precise but the small difference between the two weighings has a much larger percentage error.

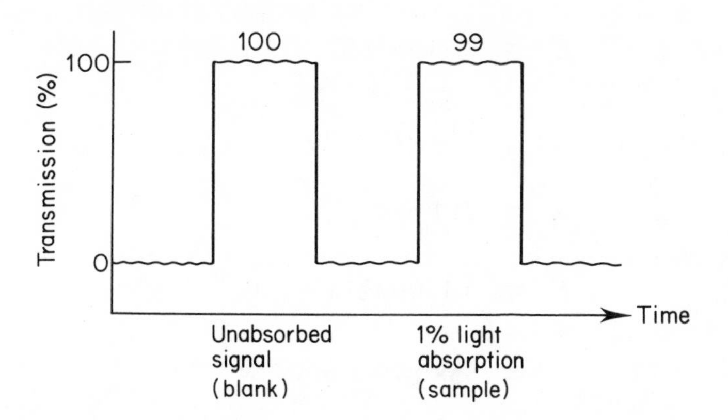

Fig. 2.2a. *The effect of absorption on the detector signal*

Π Suitable analysis wavelengths for caesium and zinc are 852 nm and 214 nm respectively. Calculate the ratios N_1/N_0 at a temperature of 3000 K, and decide for each element whether analysis by AAS or AES would be the more suitable.

(g_1/g_0 is 2 for caesium, 3 for zinc. Temperatures around 3000 K can be attained with nitrous oxide-acetlylene (ethyne) flames.)

Analysis by AES is more suitable for Cs ($N_1/N_0 = 7.19 \times 10^{-3}$), but AAS is necessary for Zn ($N_1/N_0 = 5.57 \times 10^{-10}$).

For this calculation we first need to work out the energy gap from the wavelength of the spectral line. Using Eq. 2.2

$$E = hcN_A/\lambda$$

We saw in Section 2.2 that $hcN_A = 0.1196$ J m mol^{-1} (for rough calculations the approximation $hcN_A = 0.12$ is useful).

$$E = 0.1196\ \lambda \text{ J mol}^{-1} \ (\lambda \text{ in metres})$$

For Cs, $\lambda = 852$ nm

so that $$E = 0.1196/852 \times 10^{-9}$$

$$= 140\ 376 \text{ J mol}^{-1}$$

For Zn, $$\lambda = 214 \text{ nm}$$

so that $$E = 0.1196/214 \times 10^{-9}$$

$$= 558\ 879 \text{ J mol}^{-1}$$

At 3000 K, $RT = 8.314 \times 3000 = 24\ 942 \text{ J mol}^{-1}$, so we can calculate the ratio N_1/N_0 using Eq. 2.1.

For Cs, $$N_1/N_0 = 2 \exp(-140\ 380/24942)$$

$$= 7.19 \times 10^{-3}$$

For Zn, $$N_1/N_0 = 3 \exp(-558\ 880/24\ 942)$$

$$= 5.57 \times 10^{-10}$$

The excited state population for Cs is over ten million times that for Zn. We would expect atomic emission spectroscopy to be the technique of choice for Cs but this would be unsatisfactory for Zn for which atomic absorption spectroscopy would be suitable.

The values of the ratios N_1/N_0 are plotted graphically in Fig. 2.2b for four elements. The spectral lines used vary from the caesium line in the near infrared at 852 nm to the zinc line in the ultra-violet at 214 nm. The visible lines from sodium (589 nm) and calcium (423 nm) are also shown. Note that while N_1/N_0 is much higher for the long wavelength lines, the differences between the curves are smaller at higher temperatures. All the curves approach the value of g_1/g_0 asymptotically at very high temperatures. ($g_1/g_0 = 2$ for Cs, Na and 3 for Ca, Zn.)

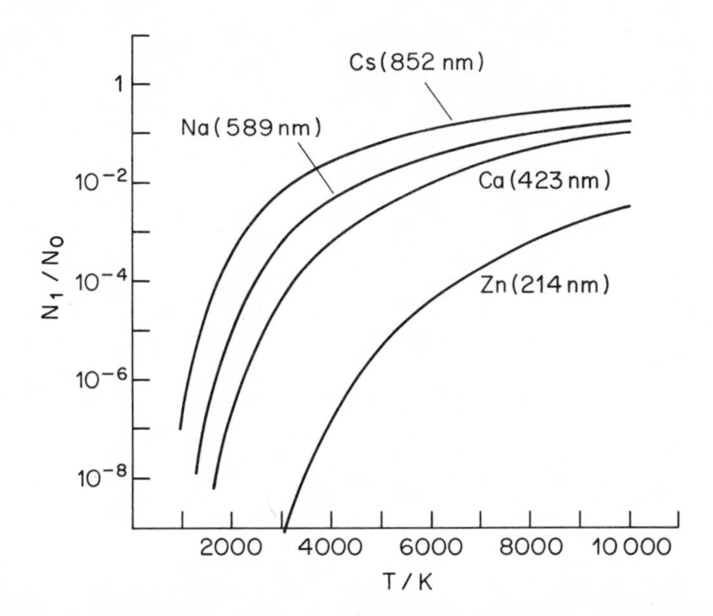

Fig. 2.2b. *The ratio N_1/N_0 as a function of temperature for Cs, Na, Ca and Zn*

SAQ 2.2a

How sensitive are atomic absorption spectroscopy and atomic emission spectroscopy to variations in the temperature of the atom cell? Flames do flicker which shows that there may well be significant temperature fluctuations. To investigate this sensitivity calculate the effect of a 10 K variation in flame temperature on N_1/N_0, for sodium atoms, by calculating N_1/N_0 at 2000 K and at 2010 K.

Use the results of these calculation to estimate the effect of the 10 K temperature rise on

(*i*) I_e, and

(*ii*) I_a.

(For sodium atoms $E = 203\ 000$ J mol^{-1} and $g_1/g_0 = 2$.)

SAQ 2.2a

2.3. SPECTRAL LINE WIDTHS

We shall see in this Section that atomic absorption and atomic emission lines are very sharp, that is, each line is spread over a very narrow range of wavelengths. To illustrate this, part of the atomic emission spectrum of silicon is shown in Fig. 2.3a. The scale is given in wavelengths and wavenumbers since both units are in common use.

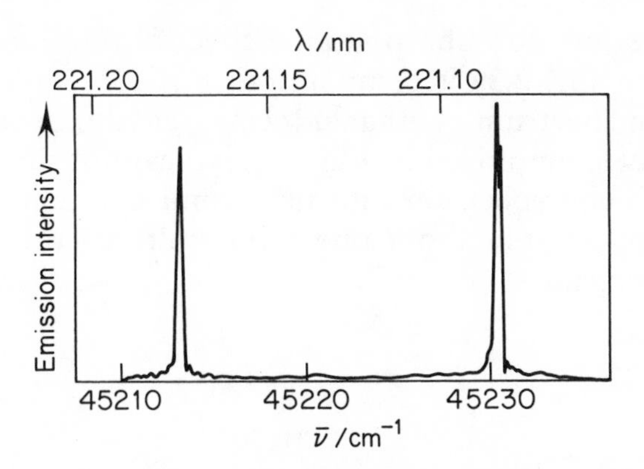

Fig. 2.3a. *Part of the atomic emission spectrum of silicon (taken from Anal. Proc., (1985), 22, 64)*

As we shall be considering the major factors influencing the magnitude of the line widths, it is necessary to be able to define the line width quantitatively. By line width we mean the difference in wavelength between the two sides of the spectral line at half the peak height. The symbol $\Delta\lambda_{1/2}$ is used for the line width defined in this way, and the definition is illustrated in Fig. 2.3b.

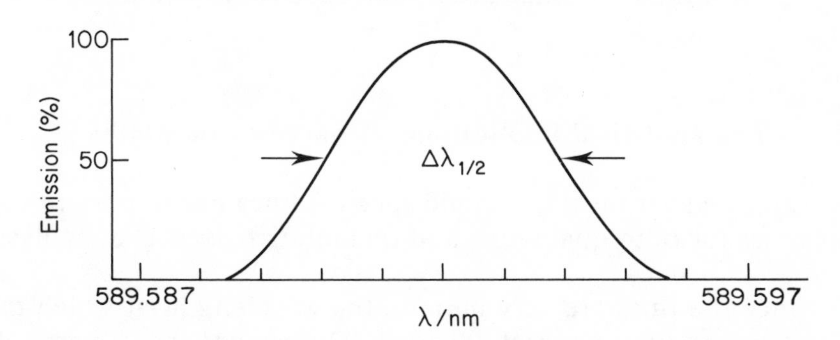

Fig. 2.3b. *Illustration of the definition of line width,* $\Delta\lambda_{1/2}$

Π Can you work out the line width from Fig. 2.3b?

The wavelengths at half the peak height are 589.590 nm and 589.594 nm, so to the nearest 0.001 nm the line width is 0.004 nm.

To emphasize just how sharp atomic spectral lines are, the silicon spectrum from Fig. 2.3a is compared in Fig. 2.3c with the molecular absorption spectrum of ethanal (acetaldehyde). Note that on the wavelength scale employed in Fig. 2.3c we cannot show the resolution of the silicon spectrum into its several components since the line separation is only about one thousandth of the width of the molecular spectrum.

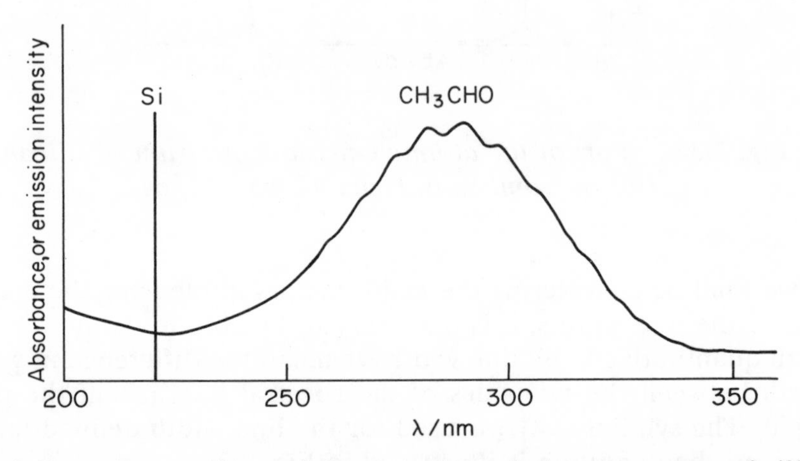

Fig. 2.3c. *Comparison of the atomic emission spectrum of silicon with the molecular absorption spectrum of ethanal*

2.3.1. The analytical implications of narrow line widths

The extreme sharpness of atomic spectral lines has important consequences for both qualitative and quantitative aspects of analysis.

(*a*) Since the lines are very narrow, the wavelengths at which they occur are very precisely defined. This means that a particular spectral line can usually be assigned to a specific element. For example, both lead and antimony have atomic absorption lines near 217 nm, but the lines are sufficiently narrow that the antimony line at 217.6 nm can be distinguished from the lead line at 217.0 nm. Even when coincidences do occur the problem can be overcome since elements usually have several useful (ie intense) spectral lines. For example antimony also absorbs at

231.1 nm and this latter wavelength could be used for analysis to avoid any possible interference due to the presence of lead in the sample.

To summarise, the sharpness and number of spectral lines of any particular element allow its presence to be detected unambiguously.

(*b*) The narrowness of atomic spectral lines also means that they are intense by comparison with molecular spectra. As shown in Fig. 2.3c the spectrum of a molecule is spread out due to the many different transitions which can occur between the ground and excited electronic states. This is because the ground state molecules can have a variety of rotational energies (typically only a few percent of molecules are in the lowest rotational energy level), and the excited electronic state formed on absorption of a photon can be in a variety of rotational and vibrational states. The result is that the intensity of the transition is spread over a wide range of energies or wavelengths. By contrast, as shown in Fig. 2.3d, since atoms cannot by their nature have rotational and vibrational energy levels, all the intensity is compressed into one transition (although more than one line may be observed in some cases).

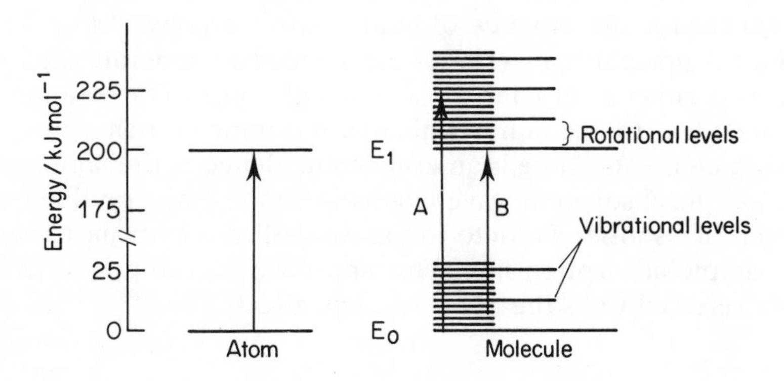

Fig. 2.3d. *Comparison of energy levels in atoms and molecules*

Π Assume that the arrows A and B in Fig. 2.3d represent the highest and lowest energy transitions in the molecular absorption spectrum. Estimate the range of energies of transitions in the molecular absorption spectrum. (The ordinate is labelled in units of kJ mol^{-1}.) Compare this with a typical atomic spectral line width of 1 J mol^{-1}.

The range of transitions energies is from 195 to 225 kJ mol^{-1}, a spread of 30,000 J mol^{-1}, (about 82 nm), and very much larger than the atomic spectral line width.

We can expect that the greater intensity of atomic transitions will lead to a greater sensitivity. This is found in practice as concentrations of around one part per million (1 ppm) can routinely be detected in atomic spectroscopy, and concentrations of around one part per billion (1 ppb, 1 in 10^9) or lower can be determined with more sophisticated procedures. For this reason atomic spectroscopy is often the method of choice for trace element determination.

To summarise, the outstanding features of atomic spectroscopy from the analytical viewpoint are the high selectivity and the high sensitivity. This allows us to both identify elements and determine them at very low concentrations.

The narrowness of atomic spectral line widths has important consequences for the practice of atomic spectroscopy. As we shall see in Part 3, special light sources are needed to generate high output intensities over a very narrow wavelength range. The solution to this problem was to use lamps which emit atomic emission lines of the analyte element. There is an exact coincidence of the lamp emission and sample absorption wavelengths. Also the emission line from the lamp is of a similar width to (or, as we shall see, even narrower than) the sample absorption line. This approach, developed by Walsh, is often referred to as the 'lock and key effect'.

2.3.2. The factors influencing line widths

Atomic spectral lines may be sharp, but the line widths are still finite. In view of the practical implications for spectrometer design it is helpful to understand which factors cause the observed line widths. The three main factors which influence line widths are considered here.

The natural line width

This is determined by the lifetime of the excited state which is formed when a photon is absorbed. An isolated atom which absorbs a photon must eventually lose the energy gained since the excited state is unstable. If the atom is isolated, the energy will be lost by emission of a photon, as shown in Fig. 2.3e.

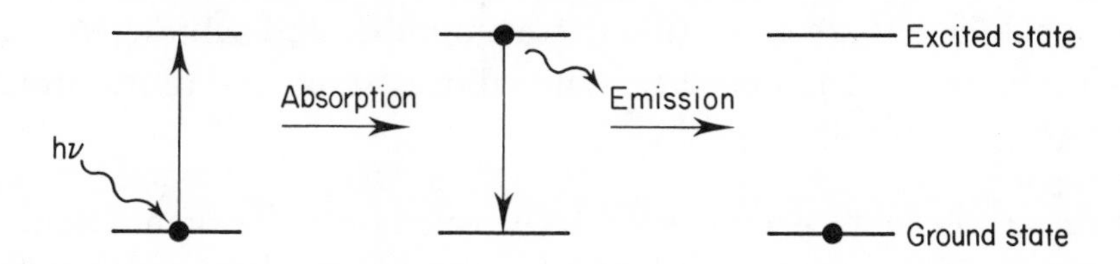

Fig. 2.3e. *The formation and decay of an excited (isolated) atom*

The absorption process is very fast (10^{-15} s), and the lifetime of the excited state is rather longer, say 10^{-9} s., but still sufficiently short that the effects of the Heisenberg Uncertainty Principle becomes appreciable. This principle states that we cannot know everything about a system exactly. In particular, we cannot know both the lifetime and the energy of an excited state with exact precision. It can be shown that this leads to an uncertainty in the wavenumber for the transition, $\Delta \bar{\nu}$, given by:

$$\Delta \bar{\nu} = 1/(2\pi\, c\, \tau)$$

$$= 5 \times 10^{-12}/\tau \qquad 2.3$$

where τ is the lifetime of the excited state.

Π What line width would you predict for an excited state lifetime of 2.5×10^{-9} s, and how does this compare with the line width measured in Fig. 2.3b?

Using Eq. 2.3 we can calculate the uncertainty in the wavenumber for the transition.

$$\Delta \bar{\nu} = 5 \times 19^{-12}/2.5 \times 10^{-9}$$

$$= 2 \times 10^{-3}\ cm^{-1}$$

The line width in Fig. 2.3b was 0.004 nm. If the wavelengths at half the peak height are converted to wavenumber and subtracted $\Delta \bar{\nu}_{1/2} = 0.115\ \text{cm}^{-1}$. The observed line width is some 500 times broader than the natural line width.

If the uncertainty is converted to wavelength units then a value of about 6×10^{-5} nm is obtained for the natural line width for the 589 nm transition for sodium atoms.

It should be clear from the results of the above calculations that the observed line widths are very much broader than the natural line widths. There must be other broadening processes which are more significant than the natural line broadening. The two most important processes are Doppler broadening and pressure broadening.

Doppler broadening

Doppler broadening is due to the rapid speeds at which the atoms in a gas move – about 1000 m s^{-1} or 2000 mph under typical conditions. Although this is a small fraction of the speed of light (3×10^8 m s^{-1}), it is still fast enough for the Doppler effect to come into play. The essence of the effect is that if a 'source' (say an excited atom in the process of emitting a photon) is moving towards a stationary 'observer' (say a detector such a a photomultiplier), then the emitted wave will appear to the observer to be bunched up. The wave will appear to have a higher frequency or a shorter wavelength. On the other hand, if the photon is moving away from the detector as the photon is emitted, the wave will appear to be stretched out, at a lower frequency or longer wavelength.

The effect is more obvious, and common to our experience, with sound waves, which travel at about 300 m s^{-1} - almost exactly a million times more slowly than light waves. As an ambulance or fire engine travelling at 30 ms^{-1} (about 60 mph) passes us we hear the pitch of the siren changing from a high pitch or shorter wavelength (source approaching) to a low pitch or longer wavelength (source receding).

In a system such as a flame, there are a great many excited atoms present. Some atoms move towards the detector as they emit photons, while others move away from the detector while emitting. As a result there will be a spread of wavelengths detected, as shown in Fig. 2.3f.

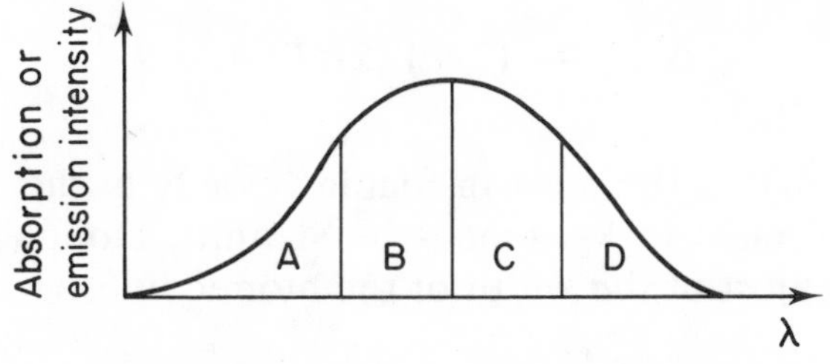

Fig. 2.3f. *Doppler broadening*

Π Label the four regions A, B, C and D of Fig. 2.3f as follows:

(*i*) atoms slowly approaching the detector;

(*ii*) atoms slowly receding from the detector;

(*iii*) atoms rapidly approaching the detector;

(*iv*) atoms rapidly receding from the detector.

The centre of the peak corresponds to molecules of negligible velocity in the direction of the detector. Regions B and C must therefore represent *slower* atoms, and A and D must represent the *faster* atoms. Regions C and D are at longer wavelengths and must represent *receding* atoms, and A and B must correspond to *approaching* atoms.

The diagram should then be labelled

A – (*iii*)

B – (*i*)

C – (*ii*)

D – (*iv*)

The width of the spectral line due to Doppler broadening is given by Eq. 2.3.

$$\Delta \lambda_{\frac{1}{2}} = (2/c)\,(2RT/A_r)^{\frac{1}{2}}\,\lambda \qquad 2.3$$

where $(2RT/A_r)^{\frac{1}{2}}$ is the most probable velocity of the atoms and A_r is the atomic mass in kg, that is in SI units. Doppler broadening accounts for most of the width of the atomic lines.

Pressure broadening

Atoms in the atom cell are not isolated as we assumed in our discussion of the natural line width. The atoms undergo frequent collisions with other atoms and molecules in the surrounding gas. Excited atoms can lose their excess energy by energy transfer during collisions. As a result the lifetime of the excited atom is very short leading to line broadening. There is no simple equation for the calculation of the resulting line width, which depends on many factors - the temperature, pressure and nature of the surrounding gas.

Π The main influence on pressure broadening is, of course, pressure. Do you think that pressure broadening will be more important for:

(*i*) a hollow cathode lamp operating at a pressure of about 0.01 atmospheres (1000 N m^{-2}); or

(*ii*) a flame operating at atmospheric pressure?

The frequency of collisions increases with concentration and therefore with pressure. Line broadening must therefore also increase with pressure, and will be more significant for (*ii*).

In practice the pressure broadening of a hollow cathode lamp is negligible, but the pressure broadening in flames can be similar to the Doppler broadening.

Pressure broadening is also known as *collisional* or *Lorentzian broadening*. (In cases where pressure broadening dominates the spectral line profile differs from the profile of a Doppler-broadened line; the line shapes are said to be *Lorentzian* and *Gaussian* respectively.)

Self-reversal broadening

This effect, also known as *self-absorption broadening*, is occasionally observed for intense resonance emission lines. An example can be seen in Fig. 2.3a - the silicon emission line near 221.08 nm has a dip in the centre. This is because the cooler atoms at the edge of the atom cell are absorbing the radiation emitted by the hotter atoms in the centre of the atom cell. Because of the temperature gradient the line width for the absorption by the cooler atoms is narrower than that for the emission from the hotter atoms.

SAQ 2.3a

The spectral line width of the 589.6 nm sodium line has been measured as 0.003 nm at 1000 K and 0.005 nm at 3000 K. Do you think this observation is consistent with the line width being determined by Doppler broadening?

SAQ 2.3a

SAQ 2.3b

A hollow cathode lamp (see Part 3) used in atomic absorption spectroscopy emits atomic iron lines but also contains neon gas and emits atomic neon lines at nearby wavelengths. If the spectral lines are mainly Doppler broadened, which of the following statements is correct?

(*i*) The neon and iron line widths are about the same.

(*ii*) The iron lines are wider than the neon lines.

(*iii*) The neon lines are wider than the iron lines.

The relative atomic masses of neon and iron are 20 and 56 respectively.

SAQ 2.3b

SUMMARY AND OBJECTIVES

Summary

Observed atomic absorption and emission spectra can be related to the energy level diagrams for the atoms. The transitions are electronic, involving the movement of electrons from one atomic orbital to another. Although the energy level diagrams are complex for atoms other than hydrogen, not all possible transitions are allowed. The interaction between electron spin momentum and orbital momentum can lead to the presence of fine structure in the spectra.

The relative sensitivity of emission and absorption techniques depends on both the photon wavelength and the temperature. Emission spectroscopy is more sensitive for higher temperatures or longer wavelength photons corresponding to lower energy gaps. Absorption spectroscopy is more favourable for atomic transitions in the ultra-violet region, or at lower temperatures.

There are several factors which influence the spectral line widths. The most important ones are, in general, Doppler and pressure broadening. Doppler broadening is particularly important at higher temperatures, but makes a significant contribution to the line width in most cases. Pressure broadening can usually be neglected at sub-atmospheric pressures.

Objectives

You should now be able to:

- relate the energy level diagrams for sodium and magnesium to the absorption and emission spectra of the atoms;
- explain why not all possible transitions in atomic spectra are observed;
- explain what a resonance line is;
- calculate the relative populations of the ground and excited states, given the energy gap and the temperature;
- relate the intensities of absorption and emission spectra to the populations of the ground and excited atomic states;
- define line width, and appreciate the narrowness of atomic spectral lines;
- explain why atomic spectroscopy can be used to identify elements unambiguously, even at very low concentrations;
- describe the main factors influencing line widths in atomic spectroscopy;
- state the conditions under which the various factors influencing line widths are likely to be important.

3. Atomic Absorption Instrumentation – Optics

Overview

More than 50% of the applications of atomic spectroscopy currently use flame-atomic absorption spectrometers. To keep detail to a reasonable level the technique of flame-atomic absorption spectroscopy is described in the next two Parts of the Unit. Many of the experimental features of AAS and AES are common since both techniques can, in principle, be carried out on the same commercial instruments. In practice different instruments are usually used for AAS and AES. This is because AAS instruments normally use a flame to form the atom cell, whereas AES instruments often employ higher temperature plasma atom cells, in conjunction with higher resolution monochromators than are needed for AAS. The reasons for, and details of, these different instrumental requirements for AES are discussed in Part 7 of this Unit. The important features of an AAS spectrometer are as follows:

Lamp → Atom cell → Monochromator → Detector
Amplifier
Recorder

The important features of AAS spectrometers can be separated into two areas – the *optics* of photon generation and detection, and the *chemistry and physics* involved in generating atoms from solid and liquid samples.

The formation of atoms in flames is a complex process and will be considered separately in the next Part of the Unit.

In this Part of the Unit we will consider the optics of AAS spectrometers. The main components used in AAS - the light source, the monochromator and the detector - are described. Particular attention will be paid to the requirements of the light source, and the most widely used light source, the hollow cathode lamp will be described in detail. We will also investigate the relationship between light absorption and sample concentration.

3.1. THE RELATIONSHIP BETWEEN LIGHT ABSORPTION AND CONCENTRATION

The amount of light absorbed in the atom cell will depend on the concentration of atoms in the atom cell. If the incident light falling on the atom cell has intensity I_0, and the transmitted light reaching the detector is I_t, as indicated in Fig. 3.1a.

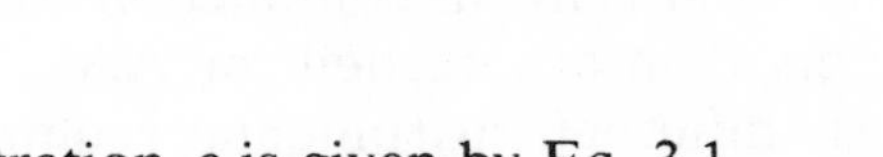

I_0 → | Sample | → I_t

Fig. 3.1a. *Absorption of incident light by the sample in the atom cell*

then the concentration c is given by Eq. 3.1.

$$\log (I_0/I_t) = \epsilon cl \qquad 3.1$$

where l is the path length through the atom cell, and ϵ is a constant (the molar absorptivity coefficient). ϵ is a measure of how intensely the atoms absorb the light, and it depends on both the nature of the atom and the wavelength. (ϵ is a maximum at the peak of the absorption curve, and falls to zero in the wings of the absorption curve). In a particular determination of a given element at a fixed wavelength ϵ is a constant, and the path length is also a constant, as this is determined by the width of the flame. The concentration then is related directly to the logarithm of the ratio I_0/I_t. A more convenient form of Eq. 3.1 can be obtained by defining the *absorbance*, A, as:

$$A = \log (I_0/I_t) \qquad 3.2$$

Therefore

$$A = \epsilon cl,$$

or

$$A \propto c \qquad 3.3$$

I_0 and I_t are usually expressed as percentages. The incident intensity I_0 is then 100%. When half of the incident light is absorbed, I_t = 50% and then

$$A = \log (100/50) = 0.301$$

When 90% of the incident light is absorbed, I_t = 10% and

$$A = \log (100/10) = 1.0$$

When 99% of the incident light is absorbed, I_t = 1% and

$$A = \log (100/1) = 2.0$$

As I_t decreases to zero than A increases to infinity. The useful practical range, however is $0 < A < 1.0$ corresponding to $100\% > I_t > 10\%$.

Some text books define A in terms of natural logarithms (ln) rather than logarithms to the base 10, which are used here to be consistent with common usage in molecular spectroscopy. If natural logarithms are used, a different absorption coefficient, k_ν, is defined, and this is related to ϵ by the relation

$$k_\nu = 2.303 \times \epsilon$$

Eq. 3.1 and 3.3 predict that A and I_t will vary with concentration as shown in Fig. 3.1b.

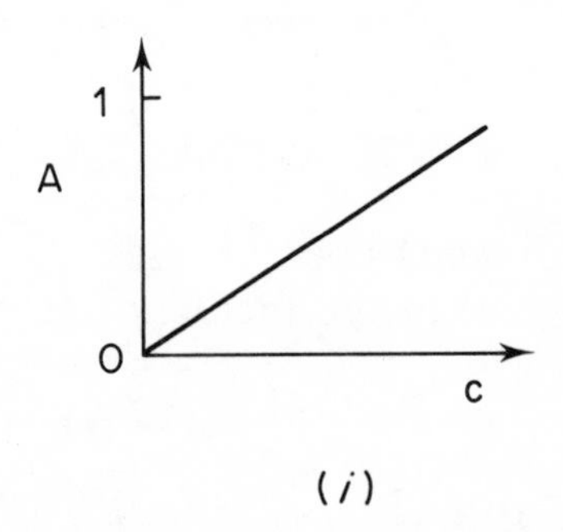

Fig. 3.1b. *(i) Variation of Absorbance with concentration*

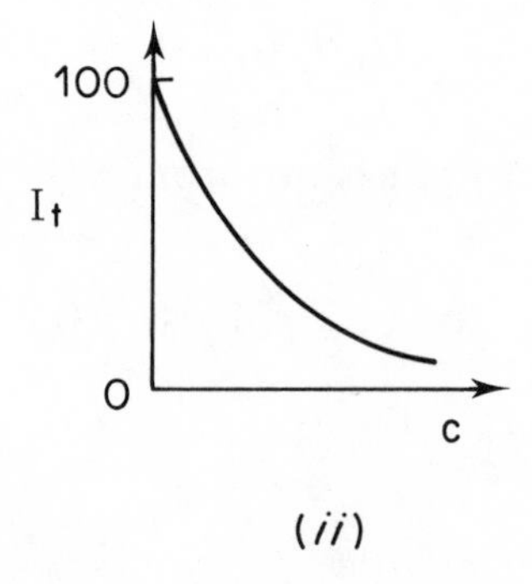

Fig. 3.1b. *(ii) Variation of intensity of transmitted light with concentration*

Π Which measurement is likely to be more useful for concentration determinations – absorbance or intensity of transmitted light?

The absorbance is far more useful than I_t since it is linearly related to the concentration, allowing the use of a plot of A *versus* c as a calibration graph. There is less error involved in interpolation of a straight line than in interpolation of a curve.

A calibration plot can be constructed by determining A for several known concentrations. There are two possible ways that the calibration plot can then be used to determine the concentration of an unknown sample. The extinction coefficient can be determined from the slope and the path length of the atom cell, and used to calculate c for an unknown from its absorbance. Alternatively the concentration of the unknown is simply read directly from the calibration plot. Although the former method is often used in molecular spectroscopy, it is less commonly used in AAS where:

(*i*) the path length is not precisely defined;

(*ii*) ϵ depends very strongly on the wavelength setting owing to the sharpness of atomic absorption lines;

(*iii*) the calibration plots are curved at higher concentrations so that the concept of a slope is not particularly useful. The curvature arises from factors such as stray light reaching the detector and the curve is usually only linear up to about 0.7 absorbance units, as shown in Fig. 3.1c.

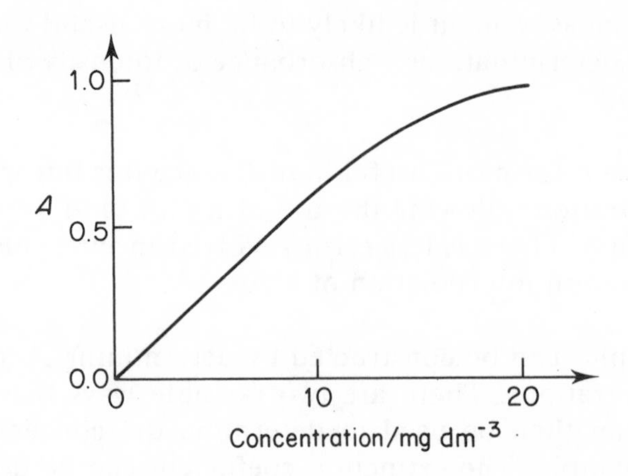

Fig. 3.1c. *Linearity of a typical calibration curve*

Π Will the precision of a determination depend on whether the absorbance of the unknown concentration is on the linear or the curved part of the plot?

The best precision is obtained using the linear portion of the curve where larger changes in A are obtained for a given increment in concentration. Check this for yourself by looking at the change in A in Fig. 3.1c on going from 0 to 10 ppm with that on going from 10 to 20 ppm.

Some modern micro-computer controlled atomic absorption spectrometers use software to "linearise" the calibration plot at higher concentrations. This allows the calibration curve to be used over a wider range of concentrations.

The units of ppm (parts per million), mg dm^{-3} or mg kg^{-1} are far more commonly used in atomic spectroscopy than the more conventional units of mol dm^{-3}. The conversion between the units is fairly straightforward and requires the use of A_r(X), the relative atomic mass of the element (X).

A 1 mol dm^{-3} solution contains A_r g of the element in 1 dm^3 of solution.

$$
\begin{aligned}
1 \text{ mol dm}^{-3} &= A_r \text{ g dm}^{-3} \\
&= 1000 \times A_r \text{ mg dm}^{-3} \\
&= 1000 \times A_r \text{ mg kg}^{-1} \text{ for dilute aqueous solutions}
\end{aligned}
$$

For example, for lead,

$$A_r \text{ (Pb)} = 207.2 \text{ g mol}^{-1},$$

therefore $1 \text{ mol dm}^{-3} = 207\,200 \text{ mg kg}^{-1}$ (or ppm)

or $1 \text{ mg kg}^{-1} = 4.83 \times 10^{-6} \text{mol dm}^{-3}$

SAQ 3.1a

The measured absorbance for a particular sample is 0.699. What percentage of light is transmitted by the sample?

SAQ 3.1b

The *characteristic concentration*, discussed in the Appendix, is a useful measure of how sensitive a technique is. It is the concentration of an element which gives rise to *1% absorption* or 99% transmission of the incident radiation. What is the corresponding value of absorbance when I_t = 99%?

SAQ 3.1c

The following values of *A* were obtained for a series of standard zinc solutions:

A	0.0	0.152	0.298	0.450	0.600	0.740	0.860	0.940
[Zn] /mg dm^{-3}	0	2	4	6	8	10	12	15

Plot the calibration curve and determine the concentration of two unknown solutions with absorbances of 0.225 and 0.900 respectively.

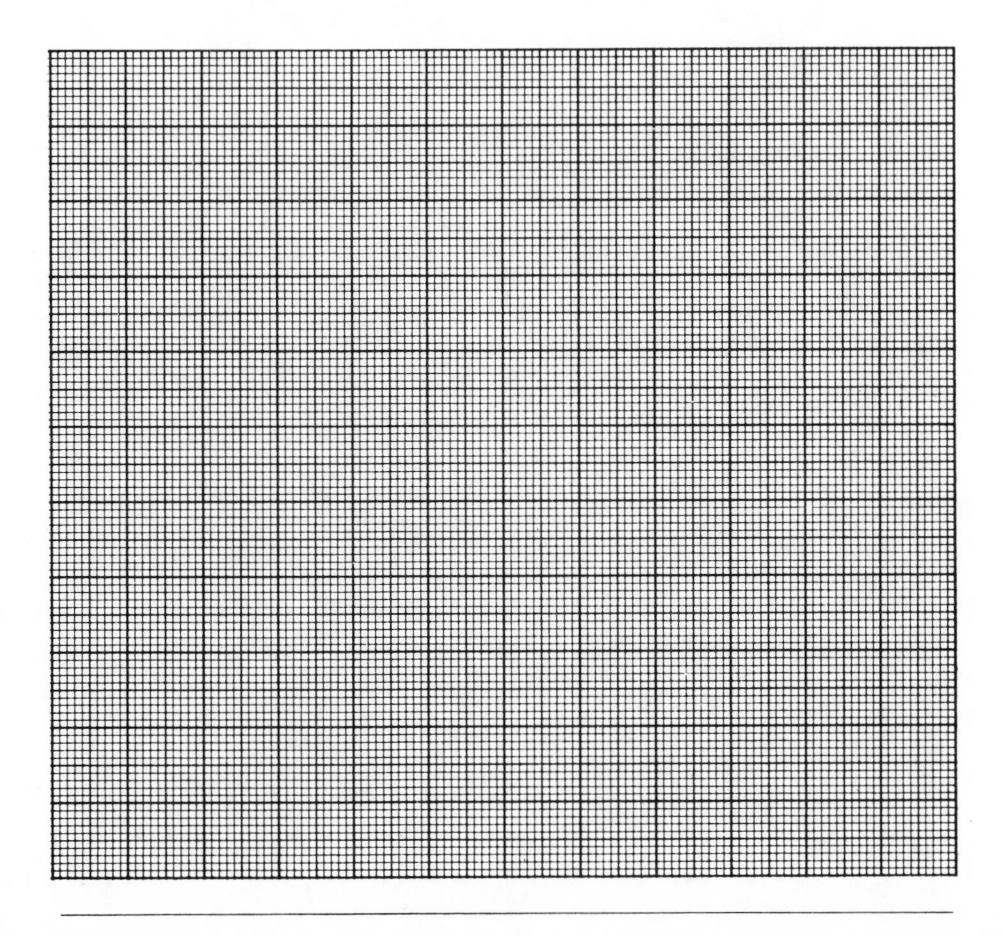

SAQ 3.1d

1.20 g of an alloy containing nickel was dissolved in hydrochloric acid and the solution made up to 100 cm^3. This solution was found by AAS to contain 150 mg kg^{-1} of nickel. What is the percentage of nickel in the original sample?

3.2. THE LIGHT SOURCE

In practice only one type of light source is widely used in atomic absorption spectrometers, the hollow cathode lamp. In this Section we shall consider the requirements of the light source for atomic absorption spectroscopy, then the hollow cathode lamp will be described in detail. Only brief reference will be made to other light sources.

3.2.1. Requirements of the source

The essential feature of atomic absorption spectroscopy is the narrowness of the absorbing spectral lines. The normal light sources used in molecular electronic spectroscopy are quartz-halogen filament lamps and deuterium or xenon arc lamps. These are called continuum sources because they emit light over a wide range of wavelengths (several hundred nm), and are used in conjunction with a monochromator to isolate the wavelength of interest. Reasonably priced monochromators can give resolutions down to 0.1 nm. To obtain narrower spectral bandwidths than 0.1 nm requires rather expensive monochromators.

Even a spectral bandpass of 0.1 nm is wider than most atomic absorption lines. We can investigate the effect of this as follows. To simplify the argument let us assume that both the lamp output and the atomic absorption line have square profiles, as shown in Fig. 3.2a. These are highly idealised profiles to make the estimation of absorbance easier, rather than the actual profiles.

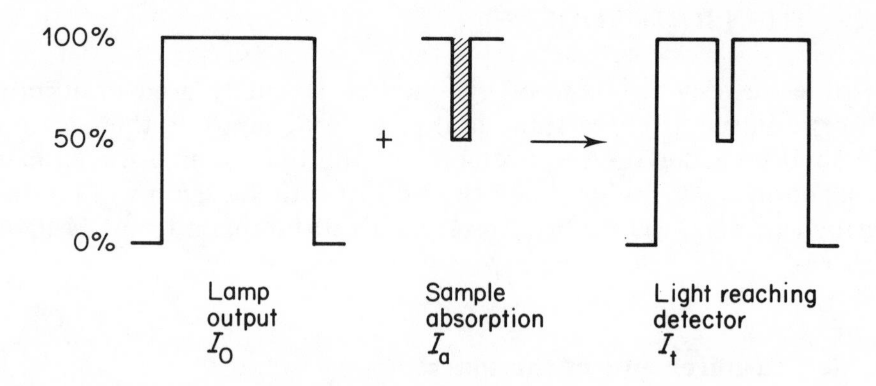

Fig. 3.2a. *Idealised profiles for lamp output and absorption line. Spectral bandpass wider than absorption*

Again for simplicity let us assume that the lamp emits a wavelength range of 0.1 nm, that is, a spectral bandpass of 0.1 nm, and that the linewidth of the absorption band is 0.01 nm. If the sample absorbs 50% of the incident radiation, then we may calculate the following quantities by relating the intensities to the areas in the diagram:

$$\text{lamp output, } I_0 = 100 \times 0.1$$

$$= 10 \text{ units}$$

$$\text{sample absorption, } I_a = 50 \times 0.01$$

$$= 0.5 \text{ units}$$

$$\text{detected light, } I_t = 9.5 \text{ units}$$

The measured absorbance, A, in this case is:

$$A = \log(I_0/I_t)$$

$$= \log(10/9.5)$$

so

$$A = 0.022$$

Now let us consider what happens when the spectral bandpass of the lamp and the linewidth of the atomic absorption are both equal to 0.01 nm. This is illustrated in Fig. 3.2b

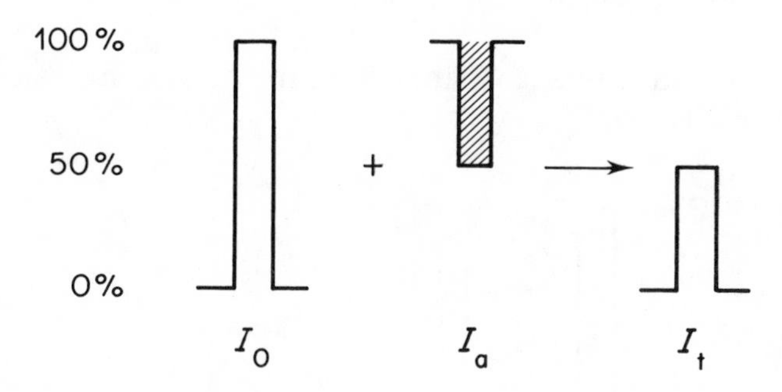

Fig. 3.2b. *Idealised profiles for lamp output and absorption line. Spectral bandpass equals absorption*

Calculating the areas as before:

$$I_0 = 100 \times 0.01$$

$$= 1 \text{ unit}$$

$$I_a = 50 \times 0.01$$

$$= 0.5 \text{ units}$$

$$I_t = 0.5 \text{ units}$$

In this case the absorbance is given by

$$A = \log(1.0/0.5)$$

so

$$A = 0.301$$

Note that this is significantly higher than the calculated absorbance in the first calculation.

Π Can you investigate the effect of the spectral bandpass of the lamp being smaller than the absorption line width? Assume that the spectral bandpass of the lamp is 0.005 nm and the absorption linewidth is 0.01 nm.

You should find that the absorbance is again 0.301. The calculation is illustrated in Fig. 3.2c.

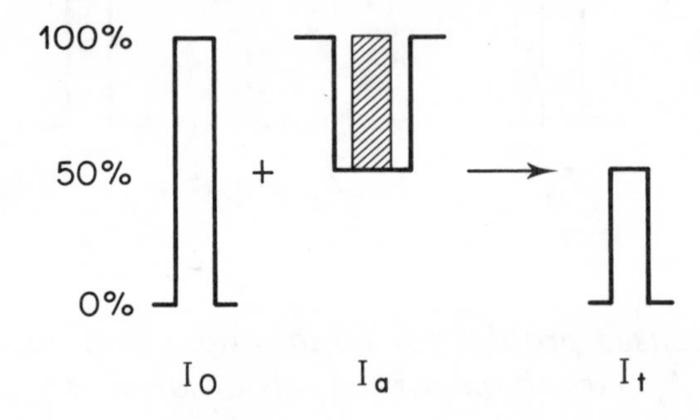

Fig. 3.2c. *Idealised profiles for lamp output and absorption line. Spectral bandpass narrower than absorption*

The important point to realise here is that only the centre of the absorption line (shaded in the diagram) can absorb light since there is no incident light at wavelengths corresponding to the wings of the absorption profile. The absorbance can then be calculated as before:

$$I_0 = 100 \times 0.005$$

$$= 0.5 \text{ units}$$

$$I_a = 50 \times 0.005$$

$$= 0.25 \text{ units (not 0.5 units)}$$

$$I_t = 0.25 \text{ units}$$

Hence the absorbance is calculated:

$$A = \log(0.5/0.25)$$

$$A = 0.301$$

The results of the three calculations are summarised in Fig. 3.2d. Some further results are included to emphasise the trends, and the values are presented in terms of the ratio, W, defined as

$$W = \text{(width of lamp spectral bandpass)/(absorption linewidth)}$$

W	10	2	1	0.5	0.1
A	0.022	0.125	0.301	0.301	0.301

Fig. 3.2d. *Calculated absorbances for different values of W*

The conclusion which can be drawn from these calculations is that *the spectral bandpass of the source should be less than or equal to the absorption linewidth*, otherwise artificially low absorbance values are obtained leading to substantial reductions in sensitivity.

In view of this conclusion we can see that the use of continuum sources with reasonably priced monochromators is not satisfactory.

3.2.2. The hollow cathode lamp

The importance of the development of the hollow cathode lamp cannot be over-emphasised, as it allowed atomic absorption spectroscopy to become the widely used technique that it is today. The hollow cathode lamp is a line source, that is it emits spectral lines rather than a continuum. The principle of the lamp is that it generates atomic emission lines using the element which is being analysed. A lamp used to analyse zinc will itself contain zinc which emits atomic spectral lines when subjected to suitable electrical discharge. This idea, developed by Walsh in the 1950's gives rise to the high selectivity of atomic absorption spectroscopy. This follows because the emission line from the lamp is

(*a*) very narrow, and

(*b*) at exactly the same wavelength as the absorbing line of the element being analysed.

(This coincidence of emission and absorption lines is referred to as the *lock and key effect*).

A schematic diagram of a hollow cathode lamp is shown in Fig. 3.2e.

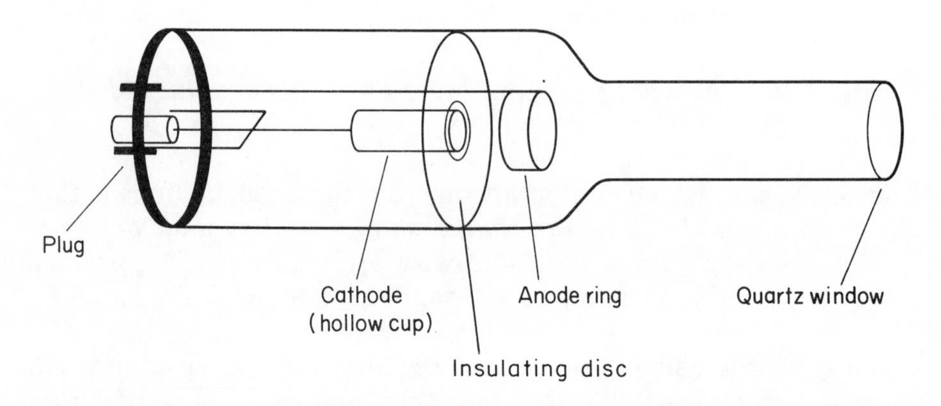

Fig. 3.2e. *The hollow cathode lamp (argon filler gas)*

The cathode is made using the metal which is to be analysed. (Where the metal is expensive or unsuitable for construction a layer of the element is deposited on a cathode made of a metal such as Fe, Al, copper or brass). The cylindrical glass tube is evacuated and filled with argon or neon at a low pressure (< 10 mmHg or $< 10^3$ N m^{-2}), and several hundred volts are applied between the two electrodes. Highly energetic electrons emitted by the cathode ionise the gas as a result of collisions:

$$Ar + e^- \rightarrow Ar^+ + 2e^-$$

The resulting ions are accelerated to the cathode, striking it with such force that metal atoms, M, are sputtered from the surface:

$$M(s) \xrightarrow{Ar^{+}(g)} M(g)$$

The metal atoms are then excited by collisions with electrons and ions:

$$M(g) \xrightarrow{e^{-},Ar^{+}} M^{*}(g)$$

The excited metal atoms then emit the characteristic atomic emission lines.

$$M^{*}(g) \rightarrow M(g) + h\nu$$

The temperature of the gas in the emission region of the discharge is only a few hundred °C.

Π Will the spectral lines emitted by the hollow cathode lamp be:

(*i*) broader than;

(*ii*) narrower than, or

(*iii*) about the same width as

the atomic absorption lines of a flame?

The hollow cathode lamp produces narrower lines than a flame ie (*ii*).

As we saw in Part 2 of the unit, the lines in the flame will be broadened by both Doppler and pressure broadening. Doppler broadening increases with temperature, and so will be smaller at the lower temperature of the hollow cathode lamp. The pressure broadening will be negligible at the low pressures of a hollow cathode lamp. The lamp emission lines then will be significantly narrower than the flame absorption lines.

The profiles of the emission and absorption lines are illustrated in Fig. 3.2f. An arrow to depict a monochromator spectral bandpass of 0.1 nm is included.

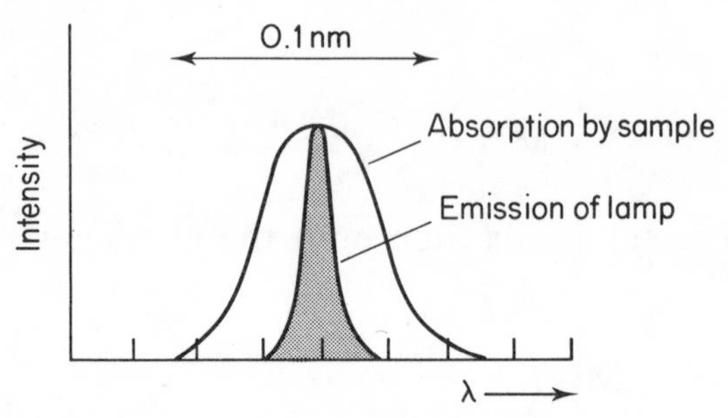

Fig. 3.2f. *Comparison of hollow cathode lamp emission line with sample absorption line*

The lamp clearly satisfies the requirement derived earlier in this Section that the source line should be equal to or narrower than the sample absorption line.

Π Is a monochromator still required when a hollow cathode lamp is used?

Yes, a monochromator is still needed since there will be several emission lines from any lamp. A monochromator is needed to select the appropriate line, but the spectral bandpass need not be too narrow - 0.1 to 1.0 nm is usually adequate.

The choice of the inert filler gas is important, as neon and argon are themselves atoms and will emit spectral lines. You have probably seen neon lights used in advertisements - the red colour of these lamps shows that neon emits light in the red part of the spectrum. The output of a calcium hollow cathode lamp in the region of the 422.7 nm calcium line is shown in Fig. 3.2g for (*i*) neon and (*ii*) argon filler gas:

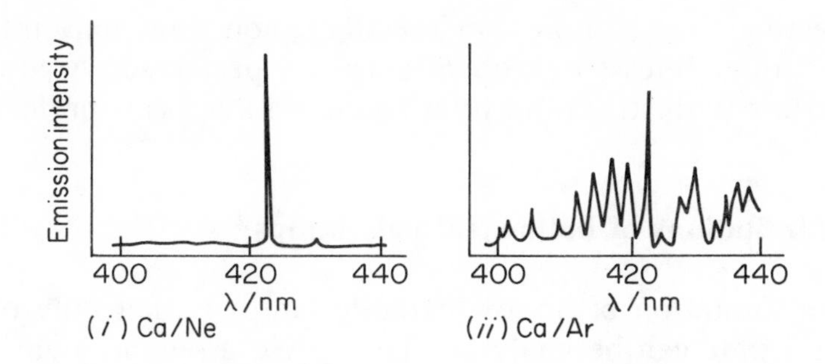

Fig. 3.2g. *Influence of filler gas on lamp output*

Π Which of the lamps in Fig. 3.2g is most suitable for calcium analysis?

The neon lamp (*i*) is more suitable since there are no neon lines close to the calcium line. With an argon filler gas difficulty may be experienced in tuning the wavelength setting of the monochromator to the correct emission line. Also if the monochromator spectral bandpass is too large, both calcium and argon lines would be detected by the photomultiplier, leading to an underestimate of absorbance for the sample, which would not absorb the argon emission passing through the monochromator.

The ranges over which neon and argon emit are shown in Fig. 3.2h.

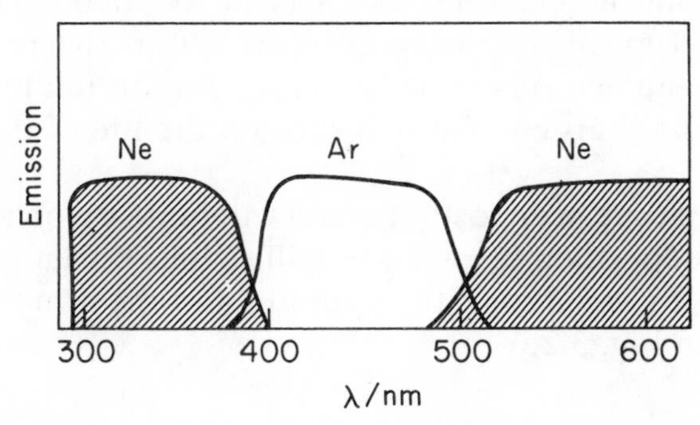

Fig. 3.2h. *Main emission regions from filler gases*

Π Which filler gas would you use for each of the following elements Nz (360 nm) and Sr (461 nm)?

By inspecting Fig. 3.2h we can see that neon lines may interfere with zirconium lines so argon filler gas is preferred. Argon lines may interfere with strontium lines so neon filler gas is preferred.

3.2.3. Limitations of hollow cathode lamps

The main limitation of hollow cathode lamps is that only one element at a time can be analysed. To analyse a second element requires changing the lamp. Even though some instruments can house several lamps on a rotating carousel, it is still necessary to align each lamp carefully to optimise the signal. The cost may also be an important factor, since although individual lamps are not expensive, the cost of building up a 'library' of many lamps can be appreciable.

Some multi-element hollow cathode lamps are available. The choice of which elements to use in any multi-element lamp is important, largely because the choice, once made, is fixed, and because the elements have to be compatible (ie not interfere with each other). The use of multi-element lamps tends to be restricted to elements which are commonly analysed together (for example a Ca/Mg lamp is available which may be useful for water analysis).

Another potential problem with hollow cathode lamps is self-absorption. This can arise from the presence of metal atoms which diffuse away from the cathode. This can be avoided by running at a low current. Typical operating currents are in the region of 10 mA, but the manufacturers usually suggest the optimum operating current. Exceeding this current will shorten the life of the lamp.

Over a period of time, a deposit of metal will build up on the walls of the lamp. Any deposit on the window will reduce the lamp intensity. this is reduced by separating the cathode and the window regions of the lamp (see Fig. 3.2e).

3.2.4. Other light sources

Continuum sources

These have already been discussed earlier in this section. The main advantage is their capacity for multi-element analysis. They are, however, rather expensive and are not, at present, commercially available.

Lasers

Tunable dye lasers are now available which can provide very high intensity spectral lines of narrow bandwidth. Although described as tunable, they are not continuously tunable in that a particular dye allows tuning only over a range of 30 – 50 nm. The dye must be changed for other ranges of wavelength. The main limitation of lasers is their high cost, which currently outweighs their advantages.

Electrodeless discharge lamps

High frequency discharges (microwave or radiofrequency) form intense light sources. These discharges are not normally used because they are less stable, and give broader lines and more self-reversal than the hollow cathode lamp discharge. There are some important areas of use for the electrodeless discharge lamp:

(*a*) a few elements for which hollow cathode lamps are difficult to operate owing to the high vapour pressure of certain volatile elements, notably As and Se;

(*b*) atomic fluorescence spectroscopy which requires very intense sources.

SAQ 3.2a

The linewidth of the 589 nm sodium line is about 0.005 nm at flame temperatures of 2000 K. Would you expect the linewidth of the resonance emission line from a hollow cathode sodium lamp operating at 500 K to be approximately:

(*i*) 0.0025 nm;

(*ii*) 0.005 nm; or

(*iii*) 0.010 nm.

SAQ 3.2b

Which of the following statements are correct?

(*i*) The measured absorbance in atomic absorption spectroscopy is not dependent on the linewidth of the source.

(*ii*) Hollow cathode lamps are less prone to self-absorption than electrodeless discharge lamps.

(*iii*) The filler gas in hollow cathode lamps can give rise to interferences.

(*iv*) Very high resolution monochromators are necessary for use with hollow cathode lamps.

(*v*) Glass windows are satisfactory for hollow cathode lamps.

3.3. THE MONOCHROMATOR

The process of separating 'white light' into its spectral components is called *dispersion*. This may be achieved by use of a prism or a grating. Filters are cheaper but pass too wide a range of wavelengths to be used on their own.

3.3.1. Prisms

The dispersion of light by a prism is illustrated in Fig. 3.3a.

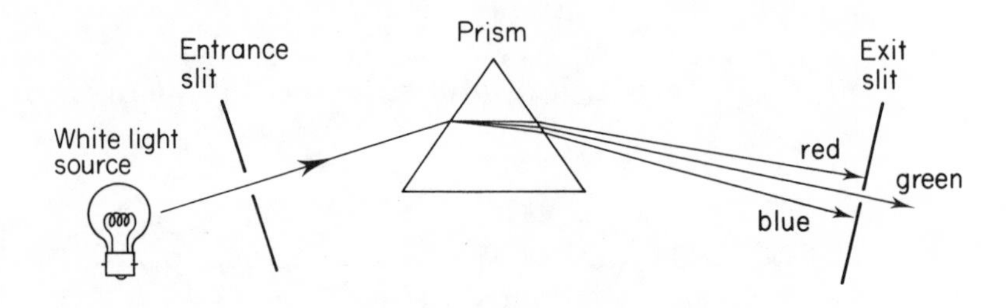

Fig. 3.3a. *A prism monochromator*

Light is refracted both on entering and on leaving the prism. The principle of a prism is that the refractive index of the prism material depends on the wavelength of the light passing through it. As the wavelength increases, the refractive index decreases. Long wavelengths are then refracted less strongly than short wavelengths. The exit slit allows a specific wavelength of light to be isolated (green light in the figure). The narrower the exit slit, the narrower the range of wavelengths which pass through it, or the more monochromatic the light is said to be.

The entrance slit is required to allow only a narrow beam of light to enter the prism. Otherwise light of several wavelengths will fall on the exit slit, arising from light beams entering the prism over a range of angles. The resolving power, R, of a monochromator measures the smallest difference in wavelength which can be separated, $\delta\lambda$, and is defined by Eq. 3.4

$$R = \lambda/\delta\lambda \qquad 3.4$$

The resolving power of a prism increases with the size of the prism, and with the rate at which the refractive index of the prism material changes with wavelength.

The transparency of optical components such as prisms in spectrometers is an important consideration. For example borosilicate glass is transparent from 310 nm to 2500 nm, while quartz is transparent from 170 nm to 2500 nm.

Π Would you use quartz or borosilicate glass for a prism monochromator for an atomic absorption spectrometer?

Both materials are satisfactory for longer wavelength analyses in the visible and near-infrared below 1000 nm. However many elements are analysed at ultra-violet wavelengths down to around 200 nm, as discussed in Part 2 of the Unit. Since borosilicate glass absorbs below 310 nm it is unsuitable and it would be necessary to use quartz.

All optical components in the spectrometer, such as lenses, need also to transmit the radiation, and they will also need to be made of quartz.

3.3.2. Diffraction gratings

Diffraction gratings can be either transmission or, more commonly, reflection gratings. Reflection gratings consist of a series of close, finely ruled lines on a thin aluminium layer deposited on optically flat glass. The lines can be ruled either mechanically, or, in the case of holographic gratings, by lasers. The transmission grating is easier to describe although the resulting equations are the same. The transmission grating consists of a series of narrow, closely spaced, parallel slits. Light passing through the slits undergoes diffraction, and for any particular wavelength, constructive interference will occur at a specific angle.

This is illustrated in Fig. 3.3b.

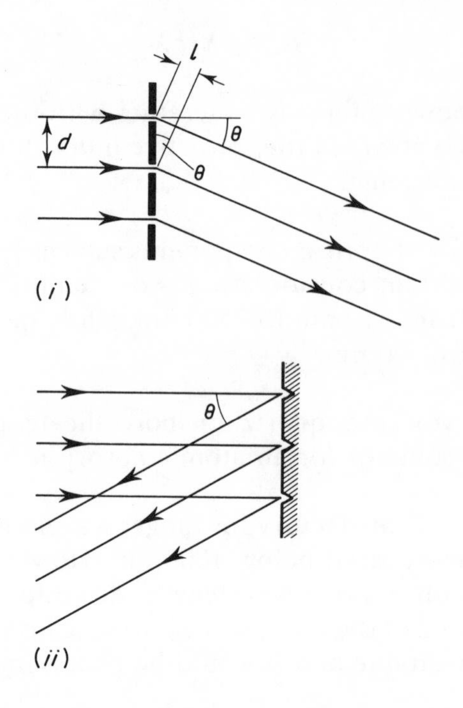

Fig. 3.3b. *(i) the transmission grating*
(ii) the reflection grating

Reinforcement of waves, or constructive interference occurs when the path difference, l, between light beams from adjacent slits is a whole number of wavelengths, n. Hence

$$l = n\lambda \qquad n = 1,2,3.. \tag{3.5}$$

(For $n=0$ all wavelengths are diffracted at an angle of $0°$ ie they are superimposed). We can evaluate l in terms of the angle of diffraction θ, by trigonometry

$$l = d \sin \theta \tag{3.6}$$

where d is the distance between adjacent slits. Eliminating l from Eq. 3.5 and 3.6

$$n\,\lambda \;=\; d \sin\theta \qquad 3.7$$

Eq. 3.7 is known as the grating equation. The value of n is called the order of diffraction. Thus $n=1$ is first order diffraction, $n=2$ is second order diffraction and so on. We may calculate the wavelength diffracted at a given angle by putting in some typical values as follows. For first order diffraction $n=1$. A typical value of d is 1×10^{-6}m (a million lines per metre, or a thousand lines per mm). For an angle of 30°:

$$1 \times \lambda \;=\; 1 \times 10^{-6} \sin 30$$

$$= 5 \times 10^{-7}\ \mathrm{m}$$

$$= 500\ \mathrm{nm}$$

The values used in the calculation are physically reasonable.

Π Will other wavelengths be diffracted at the same angle?

Yes, since there are several values of n, each corresponding to a different wavelength, which satisfy the grating equation. For $n=2$:

$$2 \times \lambda \;=\; 500\ \mathrm{nm}$$

so

$$\lambda \;=\; 250\ \mathrm{nm}$$

and there will be a series of higher order wavelengths diffracted at the same angle.

Potential problems due to the overlapping of wavelengths of different orders can be avoided by the use of filters. In the above example, shorter wavelengths can be eliminated by using an ultra-violet absorbing filter (eg glass). Another way of removing overlapping orders is to use blazed gratings. Fig. 3.3c gives a schematic representation of a blazed reflection grating.

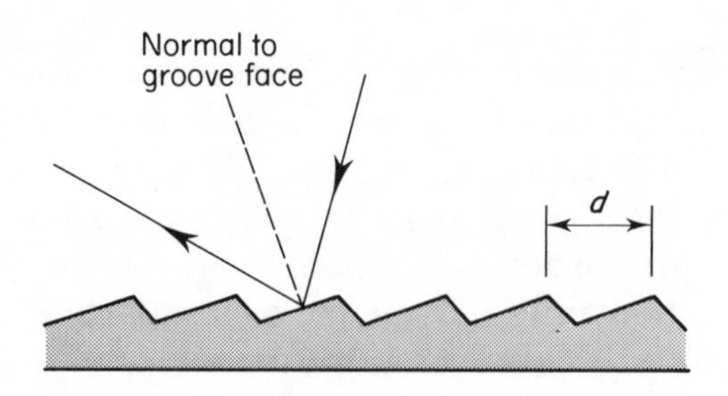

Fig. 3.3c. *A blazed reflection grating*

Reflection gratings are often blazed to reduce the second and third order reflection. (Blazed gratings can also improve output in the uv, and prevent internal reflection in the grooves.) The grooves of the grating are ruled at an angle and appear in section like a saw-tooth waveform. The diffraction is optimised when the angle of diffraction coincides with the angle of specular reflection. The optimum wavelength can be chosen by varying the angle of the saw-tooth.

The resolving power of a grating is given (approximately) by Eq. 3.8

$$R = nN \qquad 3.8$$

where N is the total number of lines in the grating. For example a grating 100 mm long having 500 lines per mm will have 50 000 lines. The resolving power for first order diffraction is then 50 000. For a wavelength of, say, 500 nm we can calculate the value of $\delta\lambda$, the smallest difference in wavelength which can be separated, from Eq. 3.4:

$$R = \lambda / \delta \lambda$$

$$= 50\,000$$

so for a wavelength of 500 nm

$$\delta\lambda = 500/50\,000$$

$$\delta\lambda = 0.01 \text{ nm}$$

Another important feature of a grating monochromator is the dispersion. In practice the reciprocal linear dispersion (RLD) is quoted. This is the wavelength interval emerging from each mm of the exit slit. The monochromators used in atomic absorption spectrometers have RLD's of 1 to 5 nm mm^{-1}. A value of 1 nm mm^{-1} means that for an exit slit 1 mm wide a spectral bandwidth of 1 nm is passed. Thus a monochromator set at 500 nm would pass light in the range 499.5 to 500.5 nm.

Π What would be the spectral bandwidth in the above example if the exit slits were set at 0.2 mm?

Since 1 mm gives a 1 nm bandwidth, an 0.2 mm slit gives a bandwidth of 0.2 nm. Thus wavelengths in the range 499.9 to 500.1 will be transmitted by the exit slits.

Spectral bandwidths in the range 0.05 to 5 nm are adequate for most atomic absorption spectrometers.

3.3.3. Prism or grating?

One of the problems with a prism is that the dispersion is wavelength-dependent. For a fixed slit width the spectral bandpass is then wavelength-dependent. In contrast gratings have, for a given order, a constant dispersion. This means that if a bandpass of 0.1 nm is obtained at 400 nm with a given exit slit width, the band pass at 500 nm will also be 0.1 nm.

Much higher resolution can be obtained with diffraction gratings, and the range of wavelengths which can be obtained from a grating is greater than with a prism.

The production of a master grating can be very expensive, as traditionally this involves ruling thousands of lines to a very high precision with a diamond. Once the master grating is made, however, cheaper replica gratings can be cast from it. More recently holographic gratings have been introduced, in which the lines are ruled by lasers.

The main limitation of gratings is the overlapping of orders, which is not a problem when using prisms.

In practice monochromators are made using gratings because of the higher resolution, wider spectral range and constant dispersion. The overlapping order problem is solved by use of a filter, or in some cases a preliminary prism to isolate a particular order is used (for example in echelle monochromators described in Part 7).

3.3.4. The grating mounting

The main requirements of the optical arrangement in the monochromator are that it should be free from distortion and have high light transmission. High transmission is achieved by using as few reflecting surfaces as possible. One of the commonly used mountings is that devised by Czerny and Turner, and outlined in Fig. 3.3d.

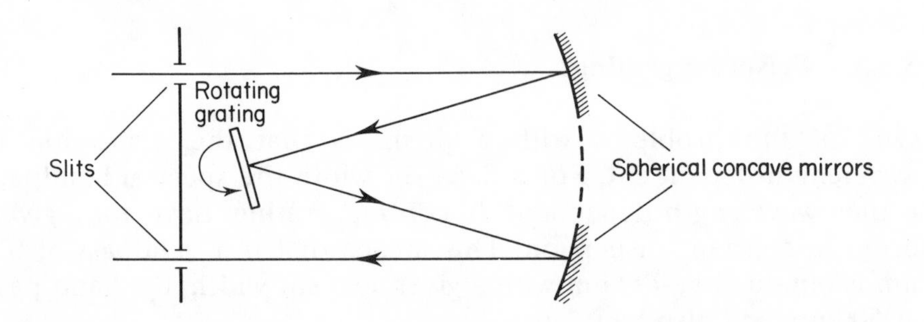

Fig. 3.3d. *The Czerny–Turner mounting*

The Czerny-Turner mounting is an economical version of the Ebert mounting, which used a single, large spherical mirror, described by the dotted line in Fig. 3.3d. Note that only three reflecting surfaces are required in this design.

In order to keep the reflecting surfaces clean the mounting is usually encased in a sealed unit, which may be evacuated.

SAQ 3.3a

Which of the following statements is correct?

(*i*) Gratings have higher resolving power than prisms.

(*ii*) Prisms have a constant (wavelength-independent) dispersion.

(*iii*) Gratings do not give overlapping orders, whereas prisms do give overlapping orders.

(*iv*) Monochromators should contain as few reflecting surfaces as possible.

3.4. THE DETECTOR

The detectors used in atomic absorption spectrometers are always photomultipliers. A photomultiplier contains a photoemissive cathode coated with an easily ionised material such as a caesium-antimony alloy. A photon falling on to the surface of the material causes an electron to be emitted provided the photon is sufficiently energetic to ionise the material, as shown in Fig. 3.4a.

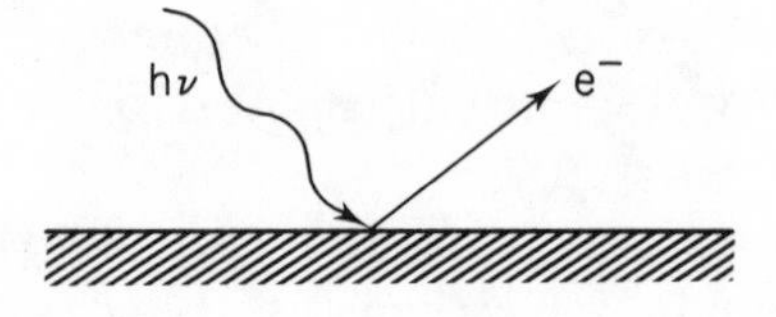

Fig. 3.4a. *Photoemission*

The surface then acts as a transducer, converting a light beam into an electrical signal.

The current produced by this process is small even though photomultiplier tubes are highly efficient (most of the incident photons generate electron emission). The signal is amplified using the process of secondary emission as illustrated in Fig. 3.4b.

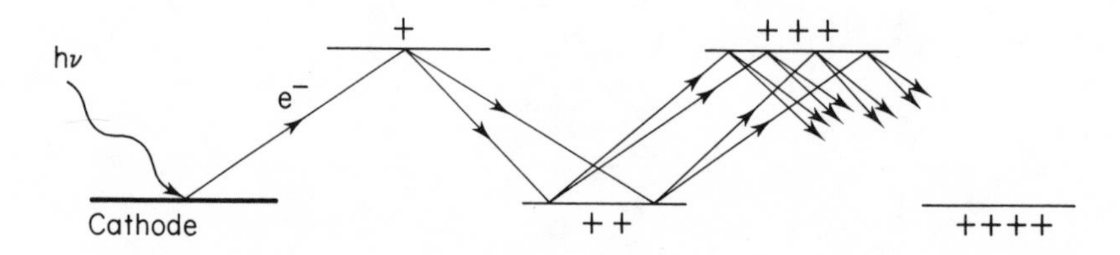

Fig. 3.4b. *A photomultiplier tube (schematic)*

The electron emitted from the cathode is accelerated to a high potential of say 100 volts. The kinetic energy aquired by the electron will be 100 eV (electron volts). This is so energetic (1 eV = 96 500 J mol^{-1}), that the electron impact on the anode causes the release of anything from 2 to 10 secondary electrons. These emitted electrons are all accelerated to a second anode which is at a potential another 100 V higher. Each electron will cause further secondary emission.

Π Typically around five secondary electrons are emitted for each incident electron. Starting with one electron, how many electrons would be produced by three successive anodes?

The first anode gives 5 secondary electrons. The second anode gives 5×5 electrons (5 for each incident electron). The third electrode gives $5 \times 5 \times 5$ or 125 electrons. The initial current has been amplified by a factor of 125.

Typical photomultipliers contain around ten anodes so that very high amplifications can be obtained.

Chemists tend to take the electronic part of a spectrometer for granted, and assume that such factors as the linearity of the detector response are adequate. Fortunately, at least in the case of photomultipliers, this blind faith is justified as the photomultiplier response (the current produced per photon) is linear to much better than 0.1% except at very high light intensity.

Some typical photomultiplier responses are shown in Fig. 3.4c.

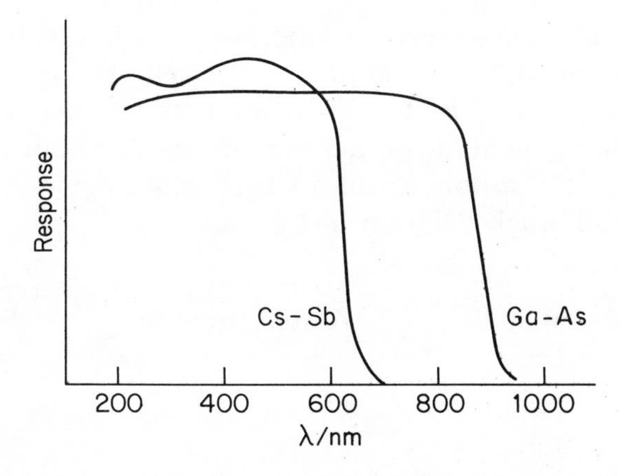

Fig. 3.4c. *Cs-Sb and Ga-As photomultiplier responses*

The long wavelength limit is determined by the photon energy required to ionise the cathode material. The short wavelength limit is determined by the material used to construct the light-transmitting window (such as quartz) of the photomultiplier.

Π Is the caesium-antimony photomultiplier, whose response was described in Fig. 3.4c, universally useful as a detector for atomic absorption spectrometers? Can it be used for studies on zinc (214 nm), sodium (589 nm) and caesium (852 nm)?

Inspection of Fig. 3.4c shows that the Cs-Sb photomultiplier does not respond to photons of wavelength above 700 nm. While being adequate for use in the visible (Na) and ultra-violet (Zn) the photomultiplier can not be used for Cs in the infra-red.

For longer wavelength studies trialkali cathodes (Sb-Na-K-Cs) can be used up to over 800 nm, and gallium arsenide (Ga-As) can be used up to 900 nm.

Solar blind photomultipliers which respond only to wavelengths below 300 nm (they do not respond to sunlight) are sometimes used for working in the ultra-violet.

3.5. MODULATION OF THE SIGNAL

A wide range of commercial atomic absorption spectrometers is available with inevitable variations in design. Details of specific instruments can usually be found in the manufacturer's operating manuals, but some general points may be made about instrumentations. The basic components of a single beam flame-AAS system are shown in the block diagram in Fig. 3.5a.

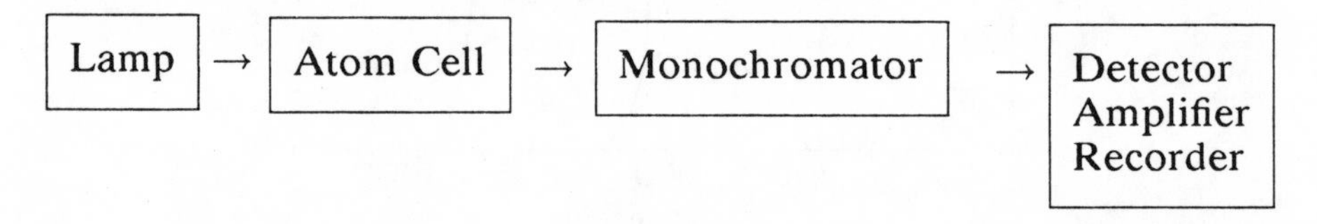

Fig. 3.5a. *The basic components of an atomic absorption spectrometer*

The light emitted from the source undergoes absorption in the atom cell and the resulting light intensity, I_t, is measured by the detector and amplifier. The incident intensity I_0 is the signal obtained when a blank sample, ie the solvent containing zero concentration of the analyte element, is introduced into the atom cell. The readout is then adjusted to give 100% transmission (zero absorbance). On introduction of the analyte solution into the atom cell, the transmitted intensity is reduced and as we have seen in Part 3.1 this reduction in intensity can be related to the concentration of the analyte.

Unfortunately the analysis of the signal is not quite so straightforward. There are two main problems.

(*i*) The system described generates a constant voltage (DC signal) at the detector during sampling. DC signals are rather prone to electrical noise, leading to poor sensitivity (or signal to noise ratios) by comparison with that obtainable for alternating current (AC) signals.

(*ii*) So far we have assumed that the only light reaching the detector comes from the lamp, but the very fact that we can see flames shows that there will also be light reaching the detector from the flame. The emission from the flame can be either molecular, from species such as C_2, CH, OH and CN, or atomic, from elements in the atom cell other than the analyte. As a result the measured light intensity at the detector may not be an exact measure of I_t, even with the use of a monochromator to remove emission at wavelengths other than the wavelength of interest.

These problems can be greatly reduced by the design of the optics and by using modulation of the light beam.

The optics are normally arranged so that the light source is focussed at the centre of the flame, and again at the entrance slit of the monochromator, as shown in Fig. 3.5b. As a result any emission from the flame, which is undesirable in an absorption experiment is minimised, since this occurs over a wide region of the flame and will not be focussed at the entrance slit of the monochromator.

When the light beam is modulated (coded), the detector is tuned to receive only the coded signals. The signal may be modulated by applying an AC supply to the lamp or by interrupting the beam using a rotating sector (chopper) in the light path. The latter method is shown in Fig. 3.5b. In both cases a beam of regularly varying intensity is produced which will generate an alternating signal at the detector. As this signal is at a fixed frequency, then by using an AC amplifier which is tuned, that is only amplifies at the same frequency as the beam modulation, all the noise at other frequencies cn be rejected giving an improvement in signal to noise ratio. A further improvement is gained if a *phase-sensitive detector* is used. This only amplifies signals which are at both the same *phase* and frequency as the modulation of the light beam.

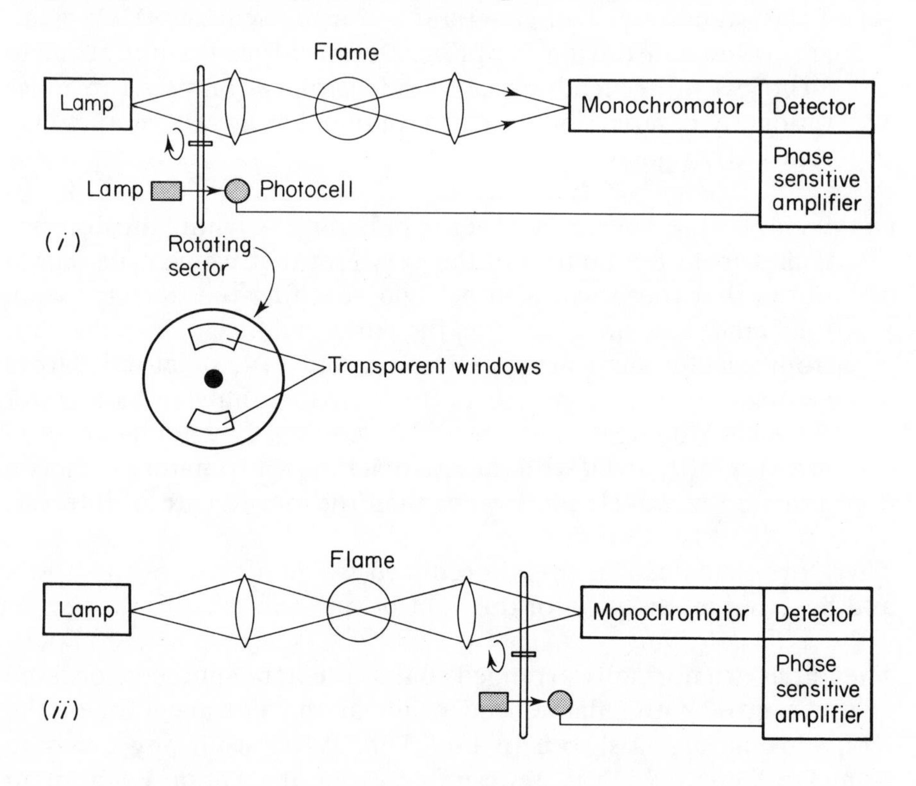

Fig. 3.5b. *Beam modulation using a rotating sector*

(i) Sector before flame
(ii) Sector after flame

In addition to improved sensitivity, there is a further advantage of modulation, in that emission signals can be eliminated.

Π One of the two optical arrangements in Fig. 3.5b will result in amplification of absorption signals but not emission signals. Can you identify which one it is?

The arrangement (*i*) is suitable for the measurement of absorption signals.

In both cases the absorption signal is modulated but in arrangement (*ii*) emission is also modulated, since the sector comes after the flame.

Thus arrangement (*i*) measures absorption signals while arrangement (*ii*) measures both absorption and emission signals.

In practice flame-AAS instruments may also use a rotating sector between the flame and the detector, so that AC signals can be obtained if the spectrometer is to be operated in the emission mode. The sector after the flame is then used in the AES mode of operation.

Double beam operation

Double beam spectrometers operate by splitting the light beam into two separate beams, one of which passes through the atom cell. The other beam is termed the reference beam, except that it does not pass through the atom cell.

The important point about double beam operation is that variations in lamp output and background absorptions are readily compensated. The sample and reference beams reach the detector alternately to give an AC difference signal.

Double beam methods are widely used in other forms of spectroscopy (uv-visible, ir etc.) but are less useful in atomic spectroscopy. The main reason for this is that the reference beam does not pass through the atom cell or flame, so that the main source of noise in AAS, the flame, is not affected by double-beam operation.

Other techniques such as *background correction* are far more useful than double beam methods, and are discussed in detail in Part 6.

SUMMARY AND OBJECTIVES

Summary

The direct proportionality between the measured absorbance and the sample concentration allows the use of calibration plots which are typically linear up to absorbances of about 0.7.

The light source in atomic absorption spectrometers is, with few exceptions, the hollow cathode lamp, which produces reasonably intense lines of exactly the same wavelength as the analyte element, and satisfies the requirement for good sensitivity of having a narrower line width than the sample absorption line (the 'lock and key' effect). Other light sources are used only in a few specific applications.

The monochromator usually contains a grating rather than a prism to disperse the light, because of the higher resolution, constant dispersion and wider spectral range of gratings. The only problem with gratings is overlapping orders, but this can be overcome by the use of blazed gratings and filters.

All atomic spectrometers use photomultipliers as the detector, because of their high sensitivity, linearity and stability. The main limitation of photomultipliers is the poor response at long wavelengths.

Double-beam operation does not compensate for the major source of noise, the flame. The formation of AC signals necessary for high signal to noise ratios is achieved in single-beam operation by modulation of the light beam. Modulation also allows absorption and emission signals to be distinguished.

Objectives

You should now be able to:

- construct a calibration curve and use the linear relationship between absorbance and concentration to calculate the concentration of an unknown from the absorbance;

- appreciate that the linear portion of the calibration curve is the most useful for analysis;

- describe the factors which limit the wavelength range for which photomultipliers are used;

- appreciate why the line width of the light source should be narrower than the absorption line of the sample, and explain why hollow cathode lamps satisfy this requirement;

- appreciate the importance of using the correct inert filler gas in hollow cathode lamps;

- describe the limitations of hollow cathode lamps and the measures taken to minimise them;

- appreciate that other sources may be more appropriate in a few cases.

- specify the relative merits of prism and grating monochromators;

- explain how photomultipliers act as transducers, converting light signals to electrical signals, and then amplifying them;

- describe the factors which limit the wavelength range for which photomultipliers are used;

- explain how modulation techniques can be used to differentiate between absorption and emission signals.

4. Atomic Absorption – from Sample to Flame

Overview

In this Part of the Unit we continue our description of the flame AAS technique started in the previous part of the unit where photon generation and detection was studied. Here the sequence involved in the general process:

Sample → Atom cell

is described. The sample will normally be in the form of a solid or liquid, and must be converted to the atomic state for analysis. The usual sequence of steps involved in this process is outlined in Fig. 4a and we shall examine each of these steps in turn:

$$\text{SAMPLE} \xrightarrow{(i)} \text{SOLUTION} \xrightarrow{(ii)} \text{AEROSOL} \xrightarrow{(iii)} \text{FLAME} \xrightarrow{(iv)} \text{ATOMS}$$

Fig. 4a. *Steps involved in atom formation*

(*i*) We will first examine the ways in which suitable solutions can be obtained from the original sample.

(*ii*) The solutions are then converted into an aerosol, and we will look at how the aerosol is formed using a *nebuliser*.

(*iii*) The aerosol is mixed with the fuel and oxidant flame gases and passes through a *burner* into the flame, and we will look at the main requirements of the burner.

(*iv*) In the flame the aerosol is decomposed to atoms. The concentration of the sample is then determined by measuring the extent of light absorption as shown in Part 3.

We shall see that some understanding of all four processes is necessary to enable the results of AAS analyses to be interpreted with confidence.

4.1. SAMPLE PREPARATION

As we saw in Part 1, most of the elements can be analysed by AAS. The elements my be present in a wide range of concentrations, from percentage levels in the elemental analysis of compounds to sub-ppm levels in trace analysis. The elements may also be present in a wide variety of chemical and physical forms. Because of the very wide range of applications of AAS, it is not possible to give a comprehensive account here of the methods of sample preparation used. Rather we shall concentrate on the principles of some of the most common methods used.

In order to devise an analytical procedure for a particular determination, several questions should be asked.

(*i*) Is the sample solid or liquid? If liquid can it be analysed directly? If solid how can it be dissolved to form a solution suitable for analysis?

(*ii*) What is the likely concentration range in which the analysis is to be made? Once the concentration range is known a suitable absorption line can be selected for the analysis, and standard solutions for the calibration can be prepared.

(*iii*) Are there likely to be any problems in the determination, such as the presence of other interfering elements?

It will also be necessary to ask which is the most suitable type of flame for the analysis, since the method of sample preparation may not be the same for different types of flame, as we shall see in Sections 4.3 and 4.4.

Instrument manufacturers usually provide handbooks with details of spectrometer operating conditions for each element, including the recommended standard solutions, absorption lines for particular concentration ranges, and flame conditions. Some of the other elements which, if present, interfere with the analytical signals may be listed.

Some samples, such as natural waters, may be suitable for direct analysis without any further treatment. Some solution samples may require the addition of an 'ionisation suppressor', a 'protective agent', or a 'releasing agent' to minimise *interferences*. (The function of these additives will be described in this Section.)

If the element is present in a solid, the solid will need to be dissolved, or if the element is in an insoluble matrix an extraction procedure will have to be used. This is best shown by a few examples considered below.

Some other procedures which are often useful such as sample preconcentration and the use of non-aqueous solutions, are also considered.

4.1.1. Dissolution of solids

The solubility of most metals is very strongly dependent on pH. Two common types of solubility behaviour are shown in Fig. 4.1a where the logarithm of solubility is plotted against the pH of the solution.

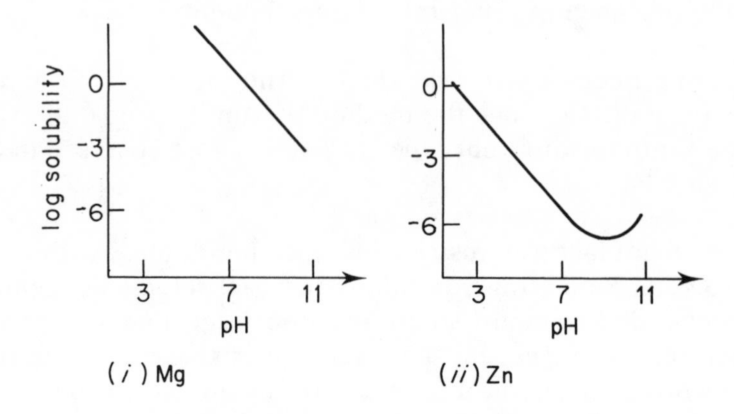

Fig. 4.1a. *Solubility-pH diagrams*

The logarithmic graph is used simply because it gives linear regions on the plots. The solubility of metals such as Mg is determined by the precipitation of the hydroxide at higher pH values. Many metals such as zinc form insoluble hydroxides ($Zn(OH)_2$) in neutral or slightly alkaline solutions, but redissolve at high pH to form complex ions (e.g. $Zn(OH)_4^{2-}$).

Π Which of the following techniques will be most generally useful for the formation of aqueous solutions from solid samples containing metallic elements?

(*i*) Dissolution in acid.

(*ii*) Dissolution in distilled water.

(*iii*) Dissolution in alkali.

The correct answer is (*i*).

Most elements dissolve in concentrated acid solutions, whereas not all elements dissolve in alkaline solutions. The solubility at moderate pH (ie in water) is often rather poor. Therefore using strong acid solution is the most generally useful method of dissolution.

Π Which of the following acids are best for dissolving metallic elements such as lead?

(*i*) nitric acid.

(*ii*) sulphuric acid.

(*iii*) ethanoic acid (acetic acid).

Nitric acid and acetic acid will give nitrates and acetates which tend to be soluble, whereas sulphuric acid is less acceptable as it may form insoluble salts (such as lead sulphate). Nitric acid, being a stronger acid than acetic acid, is more commonly used.

4.1.2. Solvent extraction

There are two situations where solvent extraction can be beneficial:

(*i*) an element may be present in too low a concentration for direct analysis;

(*ii*) other species which interfere with the determination may be present in solution.

If a suitable solvent is available in which the metal is more soluble then solvent extraction simply involves shaking the two solutions together so that the metal is extracted into the solvent. In most cases the element is initially present in aqueous solution and is extracted into an organic solvent. The organic solvent must be *immiscible* with water. The element must also be much more soluble in the organic solvent than in water. (The *partition coefficient*, which is the ratio of the equilibrium concentrations of the element in the organic solvent and in water, should be large.)

Π A particular metallic element is present in aqueous solutions at levels of about 0.5 ppm. These levels are too low for accurate analysis, although a level of 5 ppm would be adequate. If you had 100 cm^3 of the aqueous solution, how many cm^3 of solvent would you use for the solvent extraction?

The solution needs to be concentrated by factor of 5/0.5 or 10. If we assume that the extraction is 100% efficient (the partition coefficient is large), and that all the metal passes into the organic solvent, then we must use only one-tenth as much solvent. For 100 cm^3 of aqueous solution we would need 10 cm^3 of solvent.

Π Is there a simple way to check on the efficiency of the extraction?

Yes. We can carry out a second extraction on the aqueous solution and analyse the second extract for the element. If the first extraction was 100% efficient none of the element will be detected in the second extract. If the extraction is not 100% efficient however, all is not lost, because we can work out the extraction efficiency from the two analyses. This allows us to calculate the total concentration of the element.

In general an extraction will not be 100% efficient, so it is better to do more than one extraction, and to combine the extracts for analysis.

You may have some reservations about the solvent extraction. If not here is a question to think about.

Π Are metal salts usually more soluble in water or in organic solvents?

Your experience may tell you that metallic salts are usually more soluble in water, since they tend to form ionic compounds which are insoluble in non-polar solvents.

The method as we have described it then is restricted to a limited number of non-polar compounds. Fortunately there is a way that we can make the metals more soluble in the organic solvent. This is to use a *chelating agent.* A common chelating agent which you may be familiar with is EDTA (ethylene diamine tetra-acetic acid), which is a multi-dentate ligand which binds strongly to the metal ion. The equilibrium for complex formation:

$$\text{M} + \text{EDTA} \rightleftarrows \text{M (EDTA)}$$

lies well to the right for some metals, so that all the metal is present in the complexed form. The ligand wraps itself around the ion with the more hydrophobic parts of the ligand exposed to the solution. The resulting complex is then highly soluble in non-polar solvents and can be readily extracted into organic solvents.

APDC (ammonium pyrrolidine dithiocarbamate) is a very widely used chelating agent in AAS, as it is less specific than EDTA and will complex a wide range of metals. It is not usually necessary for an extraction to be highly specific – we do not have to isolate the element of interest, since most of the elements present will not cause problems of interference.

The choice of the organic solvent is important as it must:

(*i*) be immiscible with water;

(*ii*) give high solubility to the complex;

(*iii*) not give rise to problems in the flame, such as molecular absorption or emission. In the case of aromatic or halocarbon solvents this can be a significant source of error.

A solvent which satisfies these requirements and is widely used is *4-methyl-2-pentanone* (MIBK - the initials of the non-systematic name methyl isobutyl ketone.)

An organic solvent may improve the determination in other ways. For example the absorption signal for nickel in *n*-pentane may be 10 or 20 times stronger than the absorption signal for aqueous solutions of nickel. There are several reasons for these improvements. One is the increased efficiency of nebulisation (see Section 4.2).

Another reason is the increased atomisation efficiency in the flame (see Section 4.4) – this is because organic solvents will burn, tending to give higher flame temperatures than aqueous solutions which will tend to cool the flame. Also, complexes between the metal and organic ligands tend to dissociate in the flame to the metal atoms more readily than many ionic salts of the metal.

4.1.3. Ionisation suppressors

Ionisation suppressors need to be used when easily ionised elements are analysed. In practice this means:

(*i*) the alkali metals;

(*ii*) the heavier alkaline earth metals;

(*iii*) a handful of other elements (Al, B, Ti, and some rare earths).

These elements all have relatively low ionisation potentials. At flame temperatures a significant fraction of the atoms are ionised leading to a reduction of the atom concentration.

Let us consider the case of sodium analysis. In a flame at 2500 K about 20% of the sodium atoms in the flame will be ionised.

Π Will sodium atoms, Na, and sodium ions, Na^+ absorb light at the same wavelengths?

No. They will absorb at different wavelengths because there is no 3s electron in the ion, so the 3s to 3p transition giving rise to the 589 nm sodium line will not be observed. Also because of the different electron - nucleus interactions the energy levels of the ion will not be at the same energies as for the neutral atom.

We can see that the sodium absorption signal will be depressed to 80% of its value if no ionisation were present. The amount of ionisation will differ for different elements. It will depend on the ionisation potential of the element, its concentration and the temperature.

We need to find a way of suppressing the ionisation of the element of interest. The way to do this is to add a large amount of another easily ionisable element to the analyte solution. For example potassium may be used as an ionisation suppressor for the analysis of sodium or barium.

Π Will potassium be more or less ionised than sodium? The ionisation potentials are 496 kJ mol^{-1} for Na and 419 kJ mol^{-1} for potassium?

The lower ionisation potential of potassium means that it is more easily ionised.

At 2500 K about 40% of the potassium atoms are ionised – twice as many as for sodium. When more than one easily ionised element is present the ionisation equilibria can interact. For sodium and potassium:

$$Na \rightleftharpoons Na^+ + e^-$$

$$K \rightleftharpoons K^+ + e^-$$

We can use the law of mass action to investigate the way these equilibria affect each other.

Π Assume that potassium is present in much greater concentration than sodium. What will be the effect of the large concentration of electrons generated by the ionisation of potassium in equilibrium on the ionisation of sodium?

The concentration of sodium ions will be negligible in the presence of large concentrations of potassium. In case you did not predict this, let us look at the equilibrium constant for ionisation.

$$K_{ion} = [Na^+][e^-]/[Na]$$

Since K_{ion} is a constant at a given temperature, if the electron concentration is increased (from the potassium ionisation process), then $[Na^+]$ must decrease to maintain the equality in the above equation. The presence of a large electron concentration from the ionisation of potassium will suppress the sodium ionisation. Another way of looking at this effect is to think of the high electron concentration in the flame as a constraint on the sodium ionisation equilibrium which can be reduced by the equilibrium shifting to the left.

In practice quite high concentrations of the ionisation suppressor (ca 10^3ppm) are used, to make sure that the ionisation of the analyte element in the flame or atom cell is negligible.

4.1.4. Releasing or protecting agents

Whereas we use ionisation suppressants to prevent ionisation equilibria affecting the signal intensity, releasing and protecting agents are used to prevent association equilibria affecting the signal.

Association equilibria are a problem in AAS analysis when relatively stable compounds are formed in the flame, preventing complete dissociation of the sample to atoms. Compounds such as phosphates, sulphates, silicates and oxides often dissociate only partially to atoms in the flame.

An example of the effect of association equilibria is shown in Fig. 4.1b where the effect of phosphate concentration on the absorption signal for calcium atoms is illustrated. There are two regions on the curve, a linear decrease at low phosphate levels, and a saturation region where the phosphate has no further effect.

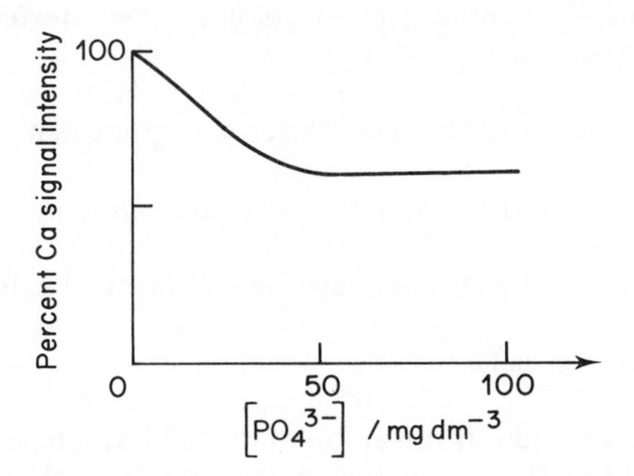

Fig. 4.1b. *Calcium signal intensity as a function of phosphate concentration*

The depression is thought to be due to the formation of relatively stable phosphate compounds.

Π Would you expect the depression of calcium signal by phosphate to depend on the flame temperature?

Yes. The effect of phosphate is reduced at higher flame temperatures. This is because the compound formed, although relatively stable, will eventually decompose if the temperature is high enough.

In addition to the reduction of the interference at higher temperatures; the interference also decreases if the measurements are made from higher in the flame (there is a longer time for the compound to decompose), and with finer aerosols from the nebuliser (see Section 4.2) due to more rapid evaporation.

Two chemical methods can be used to reduce the interference.

One method is to add a *chelating agent* (also known as a *protective agent*), such as EDTA to the solution.

Π Why do you think EDTA reduces the interference from phosphate?

(*i*) Because EDTA complexes with phosphate.

(*ii*) Because EDTA complexes with calcium.

(*iii*) Because EDTA affects the solubility of calcium.

The correct answer is (*ii*).

EDTA is an acid, and it is the anion of the acid which complexes the positive metal ion. Phosphate ions being negatively charged will not complex with the negative EDTA ion. Since the calcium is already in solution the effect of EDTA on solubility is not important.

The Ca-EDTA complex protects the Ca from the phosphate. Since EDTA is an organic molecule it will readily decompose in the flame, leaving free Ca atoms.

The second method is to add a *releasing agent* such as strontium or lanthanum, which also form complexes with the interfering anion. Competing equilibria will be set up as in ionisation suppression.

$$Ca + PO_4^{3-} \rightleftharpoons X$$

$$Sr + PO_4^{3-} \rightleftharpoons Y$$

X and Y are complexes of uncertain structure.

Π Which of the following procedures would you use for calcium analysis?

(*i*) Use about the same concentrations of Ca and Sr.

(*ii*) Use a trace of Sr.

(*iii*) Use an excess of Sr.

The correct answer is (*iii*).

We need to make all the phosphate unavailable for reaction with the Ca. To make sure that all the phosphate is complexed, a large excess of Sr should be used. If very little free phosphate is present the equilibrium for the calcium complexing with phosphate will lie well to the left.

4.1.5. Standards for calibration

We cannot easily and precisely determine the constants in Eq. 3.3 which relates the absorbance to the sample concentration ($A = \epsilon cl$) - the constant ϵ varies sharply with small changes in wavelength, and the path length l is not well defined.

In practice then we always use calibration curves. In any determination we will need a set of standard solutions of the analyte element. From the absorbance readings for the solutions of known concentration, we can draw the calibration curve. There are several important points to note on the preparation and use of standard solutions.

(*i*) A blank solution is needed to set the zero absorbance reading on the spectrometer. This should contain any additives - ionisation suppressors, protective agents etc - which have been added to the sample solutions. This is because the absorption signal is often affected by other species which may be present. The additives should also be present in the standard solutions. If little is known about the sample matrix the method of standard additions described in Section 4.1.6 is useful.

(*ii*) The samples should be freshly prepared from a concentrated stock solution, typically 1000 ppm. This is because low concentrations of a few ppm are not stable for long periods as the metal ions become adsorbed onto the surface of glass flasks. The instrument manufacturer's handbooks usually contain recipes for standard solutions, which often involve dissolving a given weight of a high purity metal or salt in acid.

(*iii*) The reagents used should be pure. This is especially important in trace element analysis where contamination from reagents may be significant. Erroneous results in trace analyses are not unknown in AAS! *AristaR* acids contain fewer impurities and are preferred to AnalaR acids for dissolving samples for trace analysis. Glass volumetric flasks should be soaked in acid then rinsed before use to remove contaminants. PTFE volumetric flasks are useful as they do not appreciably adsorb metal ions.

(*iv*) You should never extrapolate the calibration curve. There will inevitably be curvature at higher concentrations so only the linear part of the curve should be used, and the concentration range of the standards should exceed that of the samples.

(*v*) During analysis the standards should be analysed regularly as the signals may drift over a period of time. To minimise contamination effects, standards should be measured in ascending and descending order. The blank should be checked frequently.

(*vi*) The most suitable analysis wavelength should be chosen to give absorbances in the range 0.1 – 1.0.

Π The calibration curve for copper is linear up to about 10 ppm for the 324.7 nm absorption line and linear up to about 10 000 ppm for the 224.4 nm absorption line. For the 324.7 nm line the absorbance is 0.6 for the 10 ppm solution and for 224.4 nm line the absorbance is 0.6 for the 10 000 ppm solution. The noise level of the spectrometer is 0.002 on the absorbance scale. Which line would you use for the analysis of the following solutions?

(*i*) A solution containing 5 ppm Cu.

(*ii*) A solution containing 3000 ppm Cu.

(*iii*) A solution containing 50 ppm Cu.

Solution (*i*) is best analysed by the 324.7 nm line. It will be on the linear part of the calibration curve ($A = 0.6 \times 5 / 10 = 0.3$).

The absorbance for the 224.4 nm line would be too low to detect ($A = 0.6 \times 5 / 10{,}000 = 0.0003$, well below the noise level of the spectrometer).

Solution (*ii*) is best analysed by the 224.4 nm line, since it will be well beyond the linear region of the 324.7 nm line.

Solution (*iii*) requires a little more thought. It is too concentrated for analysis by the sensitive 324.7 nm line, but would give a rather low reading with the 224.4 nm line ($A = 0.6 \times 50 / 10{,}000 = 0.003$). Although just detectable, the concentration could not be measured with precision.

The simplest way round this problem is to dilute the solution tenfold (to 5 ppm), and use the 324.7 nm line. Taking eg 10.0 cm^3 and making up to 100 cm^3 in a volumetric flask introduces little error into the analysis.

One alternative would be to look for another line which is more suitable, but if other samples are already being determined by the more sensitive line, the above dilution method is simplest. Use of another line would require another calibration curve. A further method sometimes used is to rotate the burner to reduce the path length and therefore reduce the absorbance, bringing it back to the linear part of the curve. Again, another calibration curve would be needed.

4.1.6. Other calibration methods

A commonly used alternative to the use of a calibration curve is the method of *standard additions*. There are two situations where this may be used, for rapid analyses where only a few samples are to be determined, and for samples where complex matrix effects my be present.

(*i*) Suppose you have just one or two solution samples to analyse, and one standard solution made up. The usual procedure of preparing several standards and constructing a calibration curve can be rather time-consuming. A quicker procedure is to take several aliquots of the sample, then add to each, different amounts of the standard solution and make up to a constant volume and analyse.

In the following example, Fig. 4.1c, various amounts of a 100 mg dm^{-3} standard solution of the analyte element were added to 50.0 cm^3 of the solution of unknown concentration, X ppm, and the solutions were made up to 100 cm^3 in a volumetric flask.

Sample /cm^3	Standard /cm^3	Concentration /ppm	Absorbance
50.0	0.0	$X/2$	0.160
50.0	1.0	$1.0 + X/2$	0.240
50.0	2.0	$2.0 + X/2$	0.320
50.0	3.0	$3.0 + X/2$	0.400

Fig. 4.1c. *Examples of standard additions*

Π Can you work out the value of X from the data in Fig. 4.1c?

You may notice that the data are idealised to make the calculation easier. A concentration of $X/2$ ppm yields an absorbance of 0.160 while a concentration of $(2.0 + X/2)$ ppm gives exactly twice the absorbance ie 0.320. It follows then that

$$X/2 = 2.0 \text{ ppm}$$

or

$$X = 4.0 \text{ ppm}$$

The other data points are useful as checks to show the linearity of the response - each addition of the standard gives the same incremental increase in absorbance.

It is not usually so easy to carry out the calculation by inspection of the data, and the concentration is normally evaluated graphically. The above data are plotted in Fig. 4.1d.

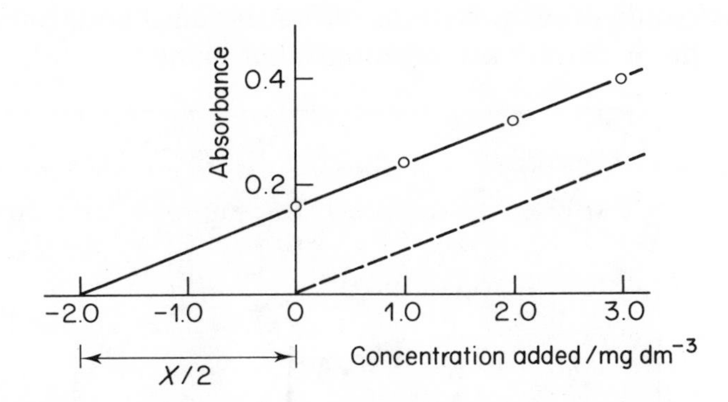

Fig. 4.1d. *Determination of unknown concentration from the standard additions plot*

The graph does not pass through the origin as in the usual calibration curves, but is effectively shifted backwards by an amount corresponding to the (diluted) sample concentration. This value is simply read off as the intercept on the concentration axis. Again we find that $X/2 = 2.0$ ppm, so that X is 4.0 ppm.

The calibration curve is shown on the graph as a dotted line. In the absence of any interference effects the standard addition curve is parallel to the calibration curve.

(*ii*) In some cases where complex solutions are analysed, interferences or matrix effects will be present.

Π In the above example will all the species in the original sample be present at the same levels in all the solutions ie the solutions containing various amounts of the standard solution?

Yes. As a consequence any matrix effect will be the same for all the solutions used to construct the calibration plot, and effects of the matrix will be compensated for.

Where it is necessary to add ionisation suppressors, protective or releasing agents, these are added to all the solutions.

There are some other variations of the standard addition method in use, but the principles are essentially the same.

SAQ 4.1a

Which of the following solvents are suitable for solvent extraction from aqueous solutions?

(*i*) Propanone

(*ii*) Methyl benzene

(*iii*) *n*-Hexane

(*iv*) Tetrachloromethane

SAQ 4.1b

Will higher temperatures increase or decrease the interferences due to:

(*i*) ionisation equilibria, and

(*ii*) association equilibria?

SAQ 4.1b

SAQ 4.1c

How many standards should you use for a calibration curve?

(*i*) 3

(*ii*) 5

(*iii*) 7

SAQ 4.1d

5.0 cm^3 aliquots of waste water were analysed for cadmium by adding various amounts of a standard solution of known cadmium concentration, and making the resulting solution up to a volume of 10.0 cm^3. Use the data below to determine the concentration of cadmium in the original water samples by means of a standard additions plot?

Added Cd /mg dm^{-3}	Absorbance
0.0	0.070
0.2	0.112
0.4	0.156
0.6	0.194

Place GRAPH PAPER (Similar to layout on p.59) here

4.2. THE NEBULISER

We have seen in the previous Section how to get the sample into a solution suitable for analysis. The next stage involves injecting the sample into a flame. We cannot simply inject the solution into the flame, because the solution will not have time to heat up enough to form atoms before the solution is carried up in the flame away from the detection system. Nebulisation converts the sample into a form more amenable to rapid atomisation.

Π Flames have *burning velocities* of around 200 cm s^{-1}. This is the speed at which the burning gases move upwards through the flame. If the height of a flame is about 5 cm, how much time does the sample spend in the flame?

Since speed = distance / time, the residence time of the sample in the flame will equal the distance divided by the speed or 5/200 = 0.025 s.

Since the sample spends only a short time in the flame we need to make sure that it will heat up as quickly as possible.

Π Which will heat up and evaporate more rapidly - large drops or small drops?

Only the surface layer molecules of solvent can evaporate. Smaller drops have a higher surface area to volume ratio than larger drops, so smaller drops will evaporate more quickly.

We need the solution to be broken up into very fine drops to form a liquid aerosol or mist before injecting the solution into the flame. This aerosol formation is the function of the nebuliser. The basic operation involves drawing up the sample at high velocity through a capillary tube to a fine jet which causes the solution to break up into tiny drops. A diagramatic representation is shown in Fig. 4.2a.

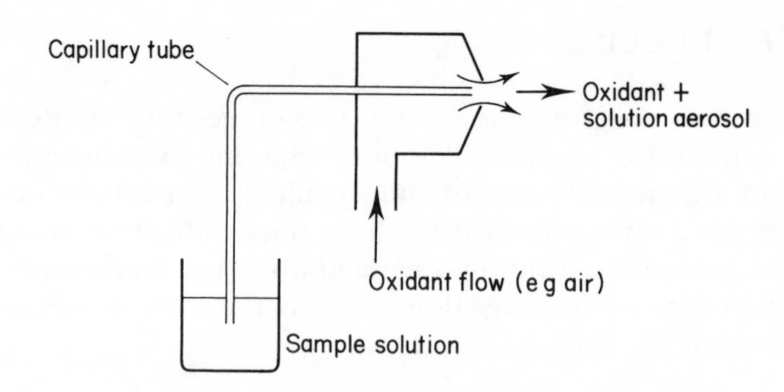

Fig. 4.2a. *Aerosol formation in the nebuliser*

The necessary suction is achieved by utilising the high flows of oxidant gas (air, N_2O or O_2) and the venturi effect. (In this the nebuliser is very much like the carburettor in a car where the high flow rate of air is used to draw up petrol through a fine jet into a mist.) In the venturi effect the high gas flow rate across the end of a capillary creates a pressure drop in the capillary. As the pressure in the capillary is below atmospheric pressure, the sample solution is forced up the capillary.

The high speed of the gas flow tends to break up the solution through turbulence as it emerges from the capillary. The aerosol formation may be encouraged in some nebuliser designs by the use of an *impact bead*. The impact bead is made of glass or alloy and placed directly after the jet, as shown in Fig. 4.2b.

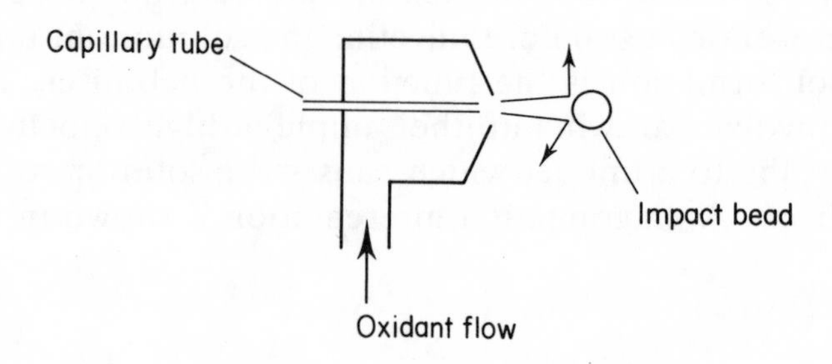

Fig. 4.2b. *The impact bead*

As well as forming the aerosol the nebuliser functions as a mixing chamber for the sample and oxidant, which are mixed at the jet, and the fuel. A fuel inlet is provided in the nebuliser as shown in the overall diagram in Fig. 4.2c. The figure also shows the presence of baffles, to filter out the larger drops, and a drain and trap to remove excess solution. Note that the trap in the drain must be filled before the flame is lit. This is to prevent flashback due to loss of pressure in the burner, and to prevent the fuel-oxidant mixture leaking into the laboratory.

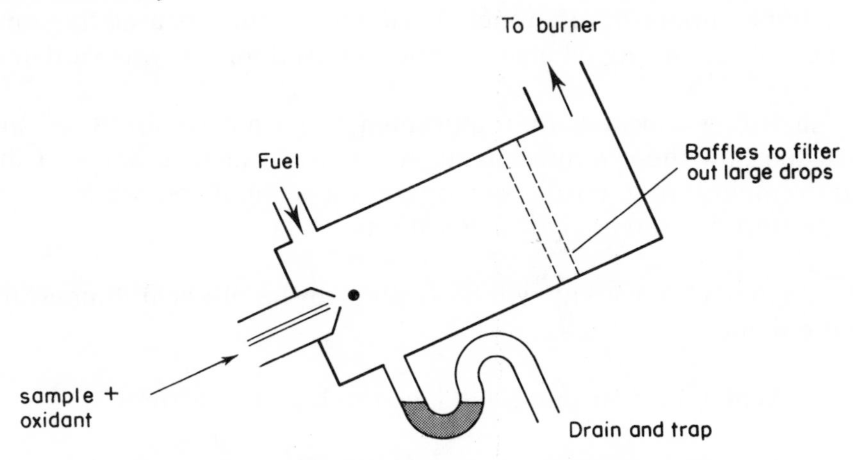

Fig. 4.2c. *Overall diagram of the nebuliser*

The efficiency of the nebuliser depends on several factors such as flow rates, surface tension, density, viscosity, saturation vapour pressure, jet size, the geometry of the jet and so on. Since different solvents will have different physical properties such as surface tension, the nebulisation efficiency depends on the solvent.

Π Water has a higher surface tension than organic solvents. Will water give more or less efficient nebulisation than organic solvents?

The lower the surface tension the easier it is to break the liquid into drops, so that organic solvents will give finer aerosols than water. The nebulisation will therefore be more efficient for organic solvents.

We have seen how the nebuliser can convert the sample solution into a form suitable for injection into the flame. In the next section we will consider the important features of the burner.

4.3. THE BURNER

Several types of flame are used in AAS, but we shall see that the basic design of the burner is essentially the same. There are, however some important differences between the burners used for some flames, so that a single burner cannot be used for all types of flame.

We shall first consider the requirements of a burner for AAS, and then see how these requirements are met in practice. Some of the requirements are to ensure safe operation of the flame, while others are to improve the analytical sensitivity.

The analytical sensitivity will be related to the physical dimensions of the flame.

Π Which do you think will give the better sensitivity?

(*i*) a long light path through the flame, or

(*ii*) a short light path through the flame.

The correct answer is (*i*).

The greater the absorbance for a given concentration the greater the sensitivity. Since absorbance is proportional to path length, ($A = \epsilon cl$, Eq. 3.3), then increasing the path length will increase the sensitivity.

Rather than using the more familiar circular type of burner (eg the bunsen burner), atomic absorption spectrometers usually use a long *slot burner* for good sensitivity. A slot burner is shown in Fig. 4.3a.

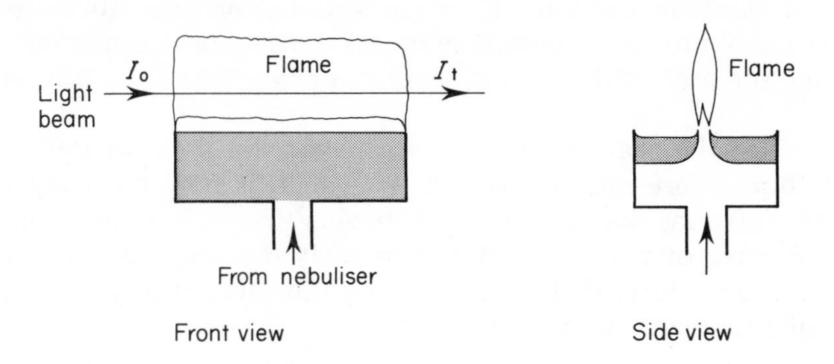

Fig. 4.3a. *The slot burner*

For the air–acetylene (air–C_2H_2) flame a slot length of 100 mm is usual. The flame height is in the region 5–10 cm. The main requirement for safe handling is that flash-back of the flame should not occur. For this the linear gas flow rate must be about three times higher than the speed at which the flame can travel (the *burning velocity*). This ensures that the flame cannot travel backwards into the nebuliser and cause an explosion. The burning velocities for different flames are in the sequence:

$$N_2O\text{-}C_2H_2 >> \text{air–}C_2H_2 >> \text{air–propane}$$

The gas volume flow rates are about the same for each flame (around 10 $dm^3\ min^{-1}$) and are largely determined by the flow of the oxidant gas. The burners must therefore have a different design in each case to give different linear flow rates. The linear flow rate of the gas, u, is determined by two factors – the volume flow rate v, and the area of the slot a, according to the relation:

$$u = v/a$$

Π Should slot burners for N_2O-C_2H_2 flames have smaller or larger slot area than for air–C_2H_2 flames?

Since v is about the same for each burner, the need for u to be larger for N_2O-C_2H_2 flames than for air–C_2H_2 flames can be met by making the area of the slot, a, smaller for the N_2O-C_2H_2 burner.

In practice the requirements for a smaller slot area for the N_2O-C_2H_2 burner are met by using a slot which is both narrower and shorter (50 mm) than the air–C_2H_2 burner. There is an unfortunate consequence of the narrower slot, namely that deposits of carbon and salts are more likely to build up on the slot during operation, affecting the flame formation.

4.3.1. The safe operation of N_2O-C_2H_2 burners

It is worth spending a little time on this, because if a N_2O-C_2H_2 burner is not operated with care, a small but rather disconcerting explosion may occur, and damage may be caused to the instrument.

(*i*) Make sure the correct burner is used.

(*ii*) Light an air–C_2H_2 flame first, and adjust it to give a smoky (fuel-rich) flame. Most spectrometers have a gas flow switch which allows the oxidant flow to be rapidly switched from air to N_2O. The important point is to maintain a high gas velocity all the time the flame is lit.

(*iii*) To switch off, turn up the fuel flow to give a smoky flame then switch back to air–C_2H_2. Turn down the fuel flow first, wait for the flame to go out then switch off the air.

More specific details for a given spectrometer are usually given in the operators manual provided by the manufacturer. Some instruments have automatic gas-boxes which carry out the above operations. If not, instruction are provided for the operation of N_2O-C_2H_2 burners which should be carefully followed.

In this Section we have considered the necessary features of a burner for use in AAS, and have seen how the slot burner can meet these requirements.

4.4. THE FLAME

In this section we look at the commonly used flames in AAS, and then at the processes involved in the conversion to an atom cell of the aerosol injected into the flame. The structure of the flame and the consequences of some of the complex flame chemistry are considered. A photograph of a typical flame is shown on the frontispiece.

4.4.1. Commonly used flames

The important flame characteristics are the temperature, and the chemical environment in the flame (ie oxidising or reducing).

The most widely used fuels are acetylene, simple alkanes, hydrogen and natural (or town) gas. The most widely used oxidants are air, oxygen and nitrous oxide. The most obvious difference between the various fuel-oxidant combinations is the temperature, but we shall see later that there are some more subtle differences in the flame chemistry which affect the choice of the combination used.

It turns out that there are two main factors which determine the flame temperature. Can you work out what these two factors are from the following questions?

Π Which flame would you expect to be hotter, oxygen–C_2H_2 or air–C_2H_2? The stoichiometric reactions are written below, taking air to be a 4:1 mixture of nitrogen to oxygen.

In oxygen:

$$C_2H_2 + 2.5\ O_2 \rightarrow 2CO_2 + H_2O$$

In air:

$$C_2H_2 + 2.5\ O_2 + 10\ N_2 \rightarrow 2CO_2 + H_2O + 10\ N_2$$

The oxygen–C_2H_2 flame will be hotter. Both reactions will produce exactly the same amount of heat per mole of C_2H_2. The heat evolved by the combustion has however to heat up all the products. In the reaction with oxygen there are 3 moles of product whereas with air there are 13 moles of product. Energy has to be used in heating up the nitrogen as well as the products of combustion, so the temperature of the air–C_2H_2 flame will be lower than the temperature of the oxygen–C_2H_2 flame.

The experimental acetylene flame temperatures are around 2450 K in air and 3100 K in oxygen, in agreement with our prediction.

Π Would you expect a highly exothermic reaction to give a lower or a higher flame temperature than a moderately exothermic reaction?

The exothermicity of the flame reaction is used to heat the flame. The more exothermic the reaction, the more heat is evolved and the higher is the flame temperature.

There are two main factors then, which determine flame temperature.

(*i*) the energy liberated in the reaction, and

(*ii*) the amount of heat absorbed by the reaction products.

The relative importance of the two factors is shown by the next question.

Π In the reaction of N_2O with C_2H_2 in flames about 1700 kJ of energy is liberated per mole of C_2H_2. Typical flame temperatures are about 2950 K. The O_2-C_2H_2 flame, described earlier liberates about 1260 kJ per mole of acetylene and gives flame temperatures around 3100 K. Which of the following statements are true?

(*i*) The nitrous oxide reaction is more exothermic than the oxygen reaction.

(*ii*) The nitrous oxide reaction generates a higher volume of products than the oxygen reaction.

(*iii*) The energy liberated is more important in determining the flame temperature than the heat absorbed by the reacting mixture.

The statement (*i*) is true.

The nitrous oxide reaction is more exothermic (1700 kJ) than the oxygen reaction (1260 kJ).

The statement (*ii*) is true.

To answer this you need to write the stoichiometric reaction:

$$C_2H_2 + 5\ N_2O \rightarrow 2\ CO_2 + H_2O + 5\ N_2$$

8 moles of product are formed for each mole of acetylene burnt in nitrous oxide. Only 3 moles of product are formed for each mole of product formed in the oxygen reaction.

The statement (*iii*) is false.

If the exothermicity were the main factor then we would expect the N_2O-C_2H_2 flame to be the hottest. But this is not the case, the N_2O-C_2H_2 flame is about 150 degrees cooler.

We can see that the flame temperature is determined more by the stoichiometry of the reaction than by the exothermicity. Another example which illustrates the point is the hydrogen–oxygen flame.

$$H_2 + 0.5\ O_2 \rightarrow H_2O$$

The heat liberated per mole of hydrogen burnt is 250 kJ. The temperature of H_2-O_2 flames can be around 2800 K, reflecting the very low volume of products formed.

In the above discussion we have assumed that the flame is stoichiometric, that is, the flame contains exactly the right amount of oxygen to concert all the fuel to carbon dioxide and water. If insufficient oxygen is present the flame is called rich, and will be much cooler, burning with a smoky, luminescent flame. If excess oxygen is present the flame is called lean. (A slight excess of oxygen can actually make the flame hotter as the flame reactions are speeded up. A large excess of oxygen will lead to a cooler flame by absorbing heat from the reaction.)

The chemistry is significantly different for lean and rich flames, as the environment is oxidising in lean flames and reducing in rich flames. It is important to use rich flames for some elements which form relatively stable oxides or hydroxides at flame temperatures (eg Al).

The most widely used flame in practice is the air–C_2H_2 mixture, with temperatures around 2450 K. For elements which require higher temperatures for efficient atomisation, the N_2O-C_2H_2 flame is usually used (3100 K). Where acetylene is not available natural gas or alkanes can be used, but the lower temperatures (1800–2200 K) give poor atomisation except for a few elements such as the alkali metals.

4.4.2. The structure of the flame

An appreciation of flame structure and chemistry is very important for two reasons.

(*i*) We need to know which part of the flame is best to use for analysis.

(*ii*) Several processes in flames can occur which affect the analysis. We have already seen that the sensitivity of emission spectroscopy is critically dependent on flame temperature, which in turn depends on flame chemistry. The other aspects of flames which are considered in this Section may either be put to good use or may lead to the unwary analyst misinterpreting the results.

The structure of a simple flame is shown in Fig. 4.4a. Four distinct regions or zones can be distinguished.

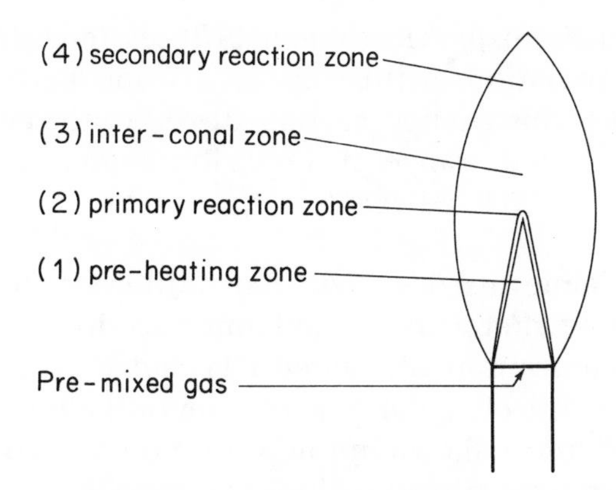

Fig. 4.4a. *Flame structure*

Region (1), the *pre-heating* zone, is the conical part of the flame which is usually observed at the bottom of the flame. In this region the temperature of the incoming gases increases very steeply as they absorb radiation from the flame.

Region (2), the *primary reaction* zone is a very thin cone, only 0.01 to 0.1 mm thick. The zone contains a great many free radicals formed in such a short time that the zone is not in local equilibrium. (There are more free radicals than you would predict from thermodynamics). There is a great deal of light emission in this zone over a wide spectral range. The tip of this zone is the hottest part of the flame. The length of the cone depends on the gas flow rate, and is carefully adjusted in practice. Too high a gas flow will lift the flame off the burner, while too low a flow may allow flash-back to occur (with a rather large bang).

Region (3), the *inter-conal* zone is in local thermal equilibrium. High temperatures are produced in this region by exothermic radical combination reactions. Much of the carbon, oxygen and hydrogen is present in the form of CO and free radical species such as H,O and OH. Less light is emitted in this region than in either the primary or secondary reaction zones.

Region (4), the *secondary reaction* zone involves the final conversion of the combustion products to CO_2 and H_2O, as the temperature falls and atmospheric oxygen diffuses into the flame. Much of the visible radiation in this region is from the oxidation of CO to CO_2 which forms excited carbon dioxide molecules which then emit light over a wide range of the visible region. Ultra-violet light is also emitted by OH radicals.

The different flames used in AAS have different colours, which reflect the different chemistry of the flames. Hydrocarbon flames contain high concentrations of excited CH and C_2 species which emit at characteristic wavelengths. A spectrum of the light emitted by the air–acetylene flame is shown in Fig. 4.4b. You may compare this with the emission from the nitrous oxide–acetylene and oxygen–hydrogen flames which are also shown.

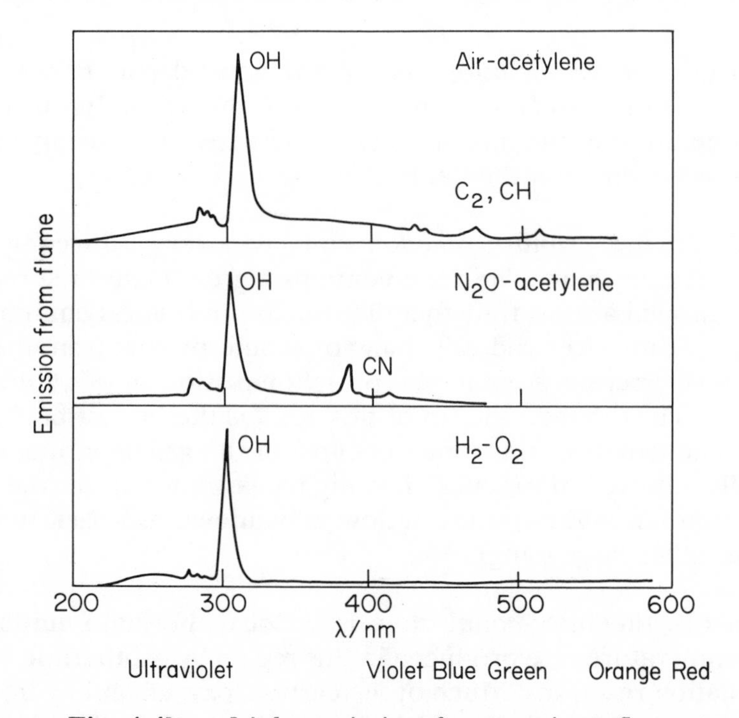

Fig. 4.4b. *Light emission from various flames*

Note that in addition to emission from C_2 in the range 450 to 600 nm, and from CH in the range 350 to 450 nm, there is also some emission from OH in the ultra-violet.

Π Would you expect the background emission from the flame to affect the atomic absorption signal intensity?

The background emission is present all the time, and the intensity of absorption, I_t, is measured as a difference signal. The signal will then be unaffected by the background emission.

In AAS, the flame emission is not too important, unless the introduction of the sample changes the nature of the emission. Changes can arise if for example, the sample contains organic matter, which will yield a higher emission from C_2 and CH, or if the sample contains such a high level of solids that a solid aerosol is formed in the flame which scatters the light beam. In these cases background correction techniques are needed, these are discussed in Part 6.

Hydrogen flames only emit from the OH radical since no carbon will be present in the flame.

Π What colour will the hydrogen–oxygen flame be?

(*i*) red

(*ii*) blue

(*iii*) green

(*iv*) colourless

The correct answer is (*iv*) – colourless.

Since the only emission, from OH, is in the ultra-violet, little visible light is emitted. As a result, the flame is almost colourless.

Nitrous oxide flames, by contrast are highly coloured. In addition to the bluish colours of the primary and secondary reaction zones (most of the emitted visible light is at the blue end of the spectrum as shown in Fig. 4.4b), the inter-conal zone is characterised by an intense red colour under fuel-rich conditions.

4.4.3. Atom formation

The sample enters the flame in the form of a very fine liquid aerosol, or mist, from the nebuliser. Several stages are involved in the formation of atoms. As an example we shall consider the fate of an aqueous solution of sodium chloride. Then we shall consider the variation of atom concentration with flame height and fuel composition. For some elements, atomisation is an efficient process, while for others the conditions required to completely atomise the sample need careful consideration. Ideally we need to completely atomise the sample to obtain maximum sensitivity and freedom from interferences.

As a fine liquid aerosol of a solution such as sodium chloride enters the flame, it passes through the high temperature gradient and the solvent rapidly evaporates, ie a solid aerosol is formed.

$$NaCl\ (aq) \rightarrow NaCl\ (s)$$

The solid then vaporises to from gaseous molecules of NaCl.

$$NaCl\ (s) \rightarrow NaCl\ (g)$$

Provided the melting and boiling points of the compound are well below the flame temperature, the rate of vaporisation is fast. Vaporisation occurs much faster than the time the atoms are in the flame. The final stage of atomisation involves the decomposition of the sodium chloride molecules to atoms.

$$NaCl\ (g) \rightarrow Na\ (g) + Cl\ (g)$$

In the above example the distance travelled through the flame before atomisation occurs is less than 1 cm in an air–acetylene flame. The profiles of sodium atom concentration against height above the burner are shown in Fig. 4.4c for both lean (L) and rich (R) flames.

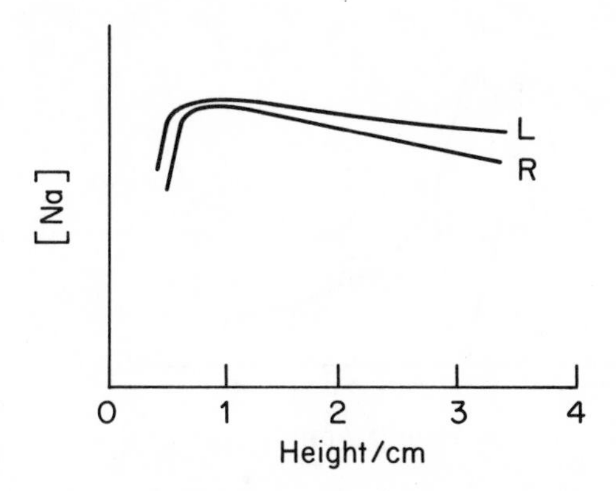

Fig. 4.4c. *Atom concentration vs flame height for Na (air–acetylene flame)*

The slow reduction of the atom concentration in the higher region of the flame is due to the cooling of the flame. This is caused by the expansion of the flame, and the dilution effect as cooler air is drawn into the flame.

Π Which is the most suitable type of flame for sodium analysis?

(*i*) Lean

(*ii*) Rich

(*iii*) Either

The correct answer is (*iii*) – either.

There is no marked dependence of atomisation on the composition of the flame gas, as can be seen from Fig. 4.4c. It does not matter too much whether we use a lean or a rich flame.

Unfortunately not all elements are so well-behaved. Elements such as aluminium and calcium which form very thermally stable oxides or hydroxides (species such as $CaOH^+$ are well-known in flames) behave differently in rich and lean flames. A rather extreme example is molybdenum, and the concentration-against-height profiles are shown in Fig. 4.4d.

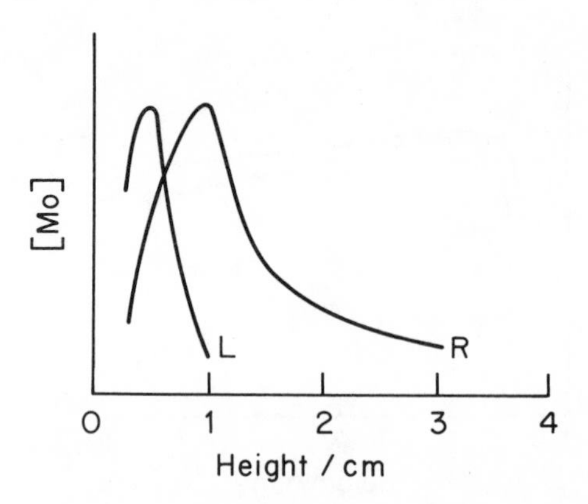

Fig. 4.4d. *Atom concentration vs flame height for Mo in lean (L) and rich (R) air–acetylene flames*

Note that, in contrast to sodium, there is a marked difference between the atom formation chemistry for flames of different composition.

Π Which is the most suitable type of flame for molybdenum analysis?

(*i*) Lean

(*ii*) Rich

(*iii*) Either

The correct answer is (*ii*) – rich.

In lean flames the higher oxygen concentrations result in loss of molybdenum atoms by the formation of stable oxide molecules. Only at the peak flame temperatures does the oxide dissociate fully. As the peak temperatures are confined to the rather narrow primary reaction zone, lean flames are unsuitable for analysis owing to the difficulty of reproducibly locating the peak atom concentration. A rich flame therefore is far more suitable for molybdenum analysis.

Note that even for rich flames the conditions for analysis of molybdenum are not as satisfactory as for sodium. In Fig. 4.4e, the concentrations of some of the species relevant to atom formation are given for rich and normal (stoichiometric) flames of air–acetylene and nitrous oxide-acetylene.

(*i*) Air–acetylene $T = 2450$ K

Species	Fuel-rich	Stoichiometric
CO	22	8
H_2	9	2
H	0.6	0.3
OH	0.1	0.3
O_2	–	0.1

(*ii*) Nitrous oxide–acetylene $T = 2950$ K

Species	Fuel-rich	Stoichiometric
CO	38	36
H_2	19	15
H	6	5
OH	–	0.4
O_2	–	0.01
CN	0.1	–

Fig. 4.4e. *Mole percentage in flames, '–' indicates very low concentration*

Π Which of the four flames described in Fig. 4.4e is most suitable for the analysis of molybdenum?

A fuel rich nitrous oxide–acetylene flame provides the best combination of high temperature and reducing conditions.

The best conditions will be high temperature and a reducing flame. At higher temperatures the oxide will be less stable and dissociate to the metal atoms. Under oxidising conditions, when appreciable concentrations of O_2 and OH are present metal oxide formation will be encouraged. Under reducing conditions where high concentrations of species such as H_2, H, CO, and CN are present, all of which tend to remove OH, O_2 and O in reactions such as

$$CO + OH \rightarrow CO_2 + H$$

thus minimising oxide formation.

The above arguments will apply to any metal which forms strong hydroxide or oxide bonds, including the alkaline earth metals, aluminium, titanium, zirconium etc. These elements form oxides and hydroxides which do not evaporate or dissociate much below 2500 K. Although fuel rich flames burn at lower temperatures than stoichiometric flames, the reducing environment is more important than the rather lower flame temperature. It is important that the analysis is confined to the observation of the inter-conal region, as the gradual diffusion of atmospheric oxygen into the higher secondary reaction zone will lead to a less reducing environment.

In low temperature flames, interferences can also arise from the incomplete dissociation of molecules other than oxides and hydroxides. A commonly observed effect is the suppression of analytical signals in the presence of halides. For example the dissociation equilibrium of sodium chloride will be pushed to the left at high concentrations of chloride ions.

$$NaCl \rightleftharpoons Na + Cl$$

At 2000 K the equilibrium constant, K_D, for the above dissociation is about 5×10^{-6}. From the expression for K_D we can calculate the ratio of free to bound sodium.

$$K_D = P_{Na} \times P_{Cl} / P_{NaCl}$$

or

$$P_{Na} / P_{NaCl} = K_D / P_{Cl}$$

For a pressure of about 1×10^{-6} atm of chlorine atoms the ratio of sodium atoms to sodium chloride molecules is 5.

$$P_{Na} / P_{NaCl} = 5 \times 10^{-6} / 1 \times 10^{-6}$$

$$= 5$$

The sodium atoms are in excess, but there is an appreciable fraction of the halide present. Not only will the signal be reduced by the presence of chlorine, but perhaps more importantly, the extent of suppression will be very sensitive to the amount of halogen present in the analyte solution, making accurate interpretation of the results difficult. As the chlorine pressure increases the signal is further reduced.

In hotter flames, the depression of signals by halide formation is not usually important, as the halide dissociation equilibria lie well to the right (see SAQ 4.4b).

SAQ 4.4a

Which of the four flame regions (1) to (4) shown in Fig. 4.4a is the best region for the analysing light beam in the spectrometer to pass through?

SAQ 4.4a

SAQ 4.4b

Can you work out the ratio of free to bound sodium at 2500 K where the dissociation constant is 2×10^{-4}? Take the pressure of chlorine to be 1×10^{-6} atm. Compare your ratio with the value of 5 calculated for 2000 K.

SAQ 4.4c

Which of the following statements are correct?

(*i*) Increasing the size of the droplets formed in the nebuliser will increase the efficiency of atomisation in the flame.

(*ii*) Air–propane flames are more likely to be prone to ionisation interferences than air–acetylene flames.

(*iii*) Nitrous oxide–acetylene flames are less likely to be prone to association interferences (such as oxide or phosphate formation) than air–acetylene flames.

(*iv*) Addition of a potassium salt to a barium solution will reduce the concentration of barium atoms in the flame.

(*v*) Addition of a strontium salt to a calcium solution will reduce the interference by phosphate on the calcium analysis.

(*vi*) Air–acetylene burners can be used for nitrous oxide–acetylene flames.

(*vii*) The sensitivity for the analysis of trace amounts of nickel in aqueous solutions can be enhanced by extracting the nickel into 4-methyl-2-pentanone (MIBK) using ammonium pyrrolidine dithiocarbamate (APDC) as a chelating agent.

SAQ 4.4c

SUMMARY AND OBJECTIVES

Summary

The process of atom formation in flames involves several stages. The sample must be prepared in a suitable form (in solution) so that a very fine aerosol can be generated in the nebuliser which will rapidly evaporate on injection through a slot burner into the flame.

If the sample is a solid it must be dissolved and/or extracted into a suitable solvent. Easily ionised elements may dissociate to ions in the flame, but this can be prevented by adding high concentrations of more easily ionised elements. Association equilibria, due to stable compound formation, can also reduce the atom concentration in the flame, but these effects can be avoided by the use of protective or releasing agents or by the use of high temperature reducing flames.

Calibration curves are normally used for analysis, and only the linear portion of the curve should be used. The method of standard additions is often useful.

Several types of flames are commonly used. The flame temperature depends on the exothermicity of the fuel–oxidant combustion reaction and on the number of moles of reaction products. The atomisation process depends on the temperature, the region and the chemical environment of the flame. The choice of flame type and stoichiometry depends on the element being analysed.

Objectives

You should now be able to:

- describe how to obtain samples in a suitable form for AAS;
- explain the use of solvent extraction for sample concentration or the elimination of interferences;
- explain how ionisation suppressors, protective (chelating) and releasing agents function;
- construct and use calibration curves;
- explain why the standard additions method can be useful, and use a standard additions plot;
- appreciate how the nebuliser generates aerosols;
- explain why slot burners are used for AAS, and why different types of flame need different burners;
- describe the differences between the various types of flame in common use;
- appreciate the different regions of flame structure and explain why the inter-conal zone is used for analysis;
- appreciate why different types of flame are required for the analysis of different elements.

5. Flameless Atom Cells

Overview

We saw in Part 4 that the flame provides a very useful way of making an atom cell, allowing us to analyse many elements down to low concentrations. But we should ask ourselves whether flame-AAS can do everything we need, or if there are some analyses for which the flame atom cell is unsatisfactory. The first question we should ask ourselves if we want to make improvements is "what are the problems which limit the technique?" Once the main problems are identified we can look at ways of overcoming them.

The main problem with flame-AAS is sensitivity. Although we have seen that it is a very sensitive technique, allowing us to determine samples at sub-ppm levels, or absolute amounts in the microgram (10^{-6} g) range, many analysts need to determine amounts down to picogram (10^{-12} g) levels. There are also some elements which are difficult to analyse at low levels by flame-AAS owing to the short wavelength of their primary resonance lines, for which alternative atomisation techniques might be usefully developed.

After a look at the main limitations of flame-AAS we shall consider how some of them can be overcome using the techniques of electrothermal atomisation, also known as furnace-AAS, and hydride generation. A few related techniques which are aimed at overcoming some of the limitations of conventional flame-AAS will also be described.

5.1. THE LIMITATIONS OF FLAME ATOMISATION

The main problem with flames is that they necessarily involve a high flow rate of sample through the flame. Although atoms are continuously formed in the flame they are also being rapidly lost as they are swept out of the observation region (the light beam) by the flow of the gases through the flame. If we know the rate of flow of the flame gases through the observation region, we can work out how long an atom spends in this region. The rate of gas flow in the most commonly used air-acetylene flames is around 200 cm s^{-1}.

Π Can you estimate the time spent in the observation region by an atom? Assume the width of the light beam is 0.2 cm.

The answer is 0.001 s. We calculate this by simply dividing the distance by the velocity, ie 0.2 cm/200 cm s^{-1} = 0.001 s.

This calculation is a little over-simplified in that the actual residence time in the observation region would be even less than a millisecond, owing to the expansion of the hot flame gases.

Another sensitivity limitation with flame atomisation is due to the nebuliser. Not all the sample drawn up into the nebuliser reaches the flame. Only the smaller drops ($< 1\ \mu m$) are used, amounting to about 10% of the sample solution.

Π If we could make an atomiser without a nebuliser, and increase the residence time of the atoms in the observation region to 1 second, by what factor would the sensitivity of the technique be increased?

The answer, at least in principle, is 10 000. A factor of 1000 comes from increasing the residence time from 0.001 to 1 s. Another factor of 10 comes from avoiding the use of a nebuliser.

So we can see that if we could lengthen the residence time of the atoms in the observation system, and avoid the use of a nebuliser, we should be able to measure much lower concentrations of atoms.

Some of the other common problems which can arise with the use of a flame are due to the form of the sample, the amount of sample which can be used, and with the use of lamps at shorter wavelengths.

(*i*) If we have a solid sample to analyse, then it must be converted to solution form either by dissolution in a suitable solvent, or, if the sample is in an insoluble matrix such as soil, by extraction. This can be very time-consuming, often taking several hours. If we analyse solids directly, a great deal of time could be saved.

(*ii*) In some cases, the amount of sample we have to analyse can be extremely small. Flame-AAS is a rather wasteful technique in these cases, since measurement involves continuously drawing the analyte solution at typically 3 cm^3 per minute into the flame,with a measurement time usually of over 10 seconds. In practice we need at least 1 cm^3 or so of solution for an analysis. If we have only a few microlitres of solution available, containing trace levels of the analyte element, we cannot use flame-AAS. The patient whose blood sample you might be analysing for suspected lead-poisoning would be much happier if you could use one drop of blood from a pin-prick, rather than remove a whole syringe full of blood.

(*iii*) The other major problem which can arise is with those elements whose primary resonance lines occur near 200 nm. This is a notoriously difficult region to use because most molecules absorb in this region, and since a flame inevitably contains molecules, a great deal of the radiation from the lamp is absorbed by the flame gases. The problem is actually worse than this partly because lamps operating near 200 nm tend to have low output intensity, due to the high energy of the photons emitted, and because of the high intensity losses for uv light at each surface in the optics. One particularly difficult element to analyse is arsenic with a primary resonance line of 193.7 nm.

In order to overcome some of the above problems it is best to avoid using a flame altogether. In some developments, however, a flame may still be used but the atom cell is separated from the flame.

5.2. ELECTROTHERMAL ATOMISATION

If we are to dispense with the flame, we have to use another method of heating the sample to a sufficiently high temperature for atomisation to occur. We can see the simplest way to heat a sample rapidly every time we switch on a light bulb. The electrically-conducting filament is heated in about a second to white heat – a few thousand degrees Kelvin. The process of heating by passing a current through an electrical conductor is an *electrothermal* one.

The most widely used electrothermal atomisation process involves placing the sample inside a hollow graphite tube, and then rapidly heating the tube, by a high electrical current (several hundred amps), to a temperature which is high enough to atomise the sample. This is shown in Fig. 5.2a.

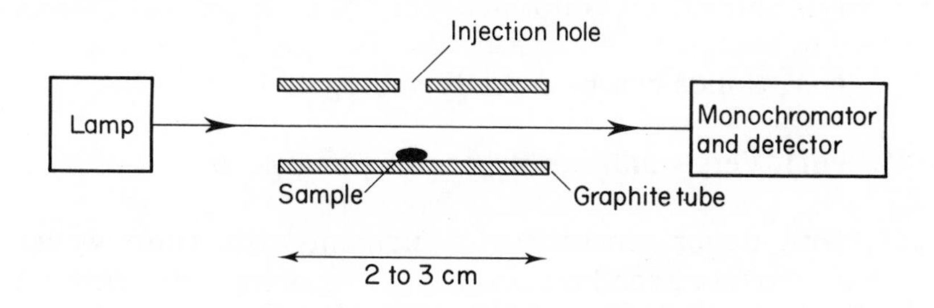

Fig. 5.2a. *The graphite furnace*

There are many variations on the design, but Fig. 5.2a illustrates the important principles. The sample (1 to 100 microlitres) is applied typically from a micropipette via the injection hole in the centre of the graphite tube. The tube is usually 20–30 mm long and 5–10 mm diameter. The atomisation takes place on rapid (1–2 s) electrothermal heating of the graphite tube. The observation light beam passes along the axis of the graphite tube. The technique is usually called graphite furnace-AAS or simply *furnace-AAS*. It is also referred to as the Massmann furnace.

The most important feature of the design is that the atoms are confined to the tube for a relatively long time (up to a second), making the technique very sensitive. The atoms are not completely confined to the furnace, because they can diffuse out through the ends of the tube and the injection hole, and there is a strong expansion of the gases in the furnace as it is heated. Also there is usually a small inert gas flow to flush the tube. This minimises the formation of refractory oxides, oxidation of the graphite and also flushes the sample from the furnace. Argon is often used as it is unreactive and doesn't absorb ultra-violet light , but nitrogen can be used except in cases where it reacts (for example titanium forms a nitride).

Since the sample is applied directly to the furnace there is no need to use a nebuliser, giving a further increase in sensitivity.

Π Which of these potential improvements with respect to flame-AAS does electrothermal atomisation give ?

(*i*) Nebuliser is not needed.

(*ii*) Solids can be directly analysed.

(*iii*) Very small samples can be analysed.

(*iv*) Better sensitivity for elements with short wavelength resonance line.

(*i*), (*ii*) and (*iii*) are all improvements which are realised in electrothermal atomisation. The case (*iv*) is not straightforward, in that while there are no flame gases to absorb short wavelength ultra-violet light, the problems which arise from background absorption of light by other species present are often more severe, even with 'furnace programming' (see below).

In practice there is a 100-fold to 1000-fold improvement in detection limit for furnace-AAS compared to flame-AAS. This is illustrated for the elements tin, calcium and zinc in Fig. 5.2b. The calculation of the relative improvement depends on how large a sample is used in furnace-AAS. This is because the flame technique measures the *concentration* of the element directly (in mg dm^{-3}), while the furnace technique measures the *amount* of the element (in mg). The amount of element must then be divided by the volume used to give the concentration. A sample volume of 20 microlitres has been assumed in the estimation of the furnace-AAS detection limits.

The figures for the elements cited are within the ranges normally found of 1 to 1000 ppb (1 part per billion = 0.001 ppm) for flame-AAS and 0.001 to 10 ppb for furnace-AAS. The definition of detection limit is given in the Appendix of this unit, but it essentially a measure of the minimum concentration which can just about be detected

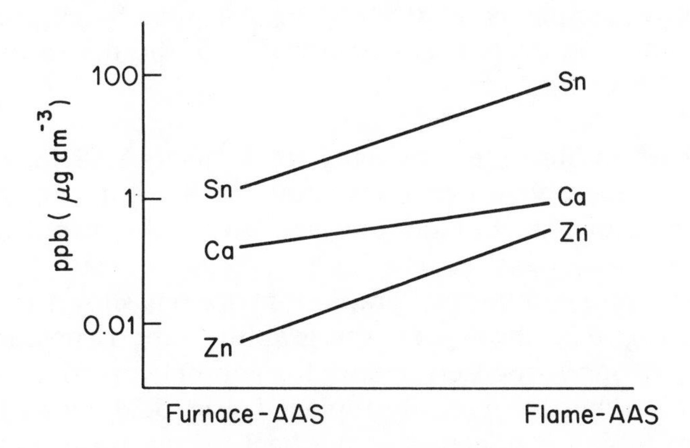

Fig. 5.2b. *Comparison of detection limits for furnace-AAS and flame-AAS*

Π See if you can use Fig. 5.2b to answer the following questions.

(*i*) Which of the elements Ca, Sn and Zn can be detected at the lowest levels for (*a*) furnace-AAS and (*b*) flame-AAS?

(*ii*) Which of the three elements benefits least from the furnace-AAS technique?

(*iii*) Estimate the detection limit using furnace-AAS for Sn if a 50 microlitre sample is used, instead of the 20 microlitre sample volume assumed, for which the detection limit is 1 ppb.

(*i*) Zn has the lowest detection limit of around 0.01 ppb for furnace-AAS. Zinc also has the lowest detection limit for flame-AAS, slightly below the level for Ca near 1 ppb.

(*ii*) The increased sensitivity for Ca is only about a factor of ten, compared with factors of about a hundred for Zn and Sn.

(*iii*) The detection limit for Sn by furnace-AAS using a 50 microlitre sample is $1 \times 20/50$ or 0.4 ppb. Since the sample amount is increased by a factor of 2.5, the detection limit is lowered by a factor of 2.5.

The increase in absolute sensitivity for furnace-AAS may be even greater than the above examples show, as these do not take into account the much smaller samples needed for the furnace.

An example of a commercial graphite furnace is shown in Fig. 5.2c and 5.2d. Fig. 5.2c shows how the graphite tube is positioned between two spring-loaded electrodes for good electrical contact, and to allow the tubes to be easily replaced. Fig. 5.2d shows the complex set of tubing connectors at the back of the furnace assembly, for water (to cool the electrodes rapidly between measurements), argon (to purge the tube), and air.

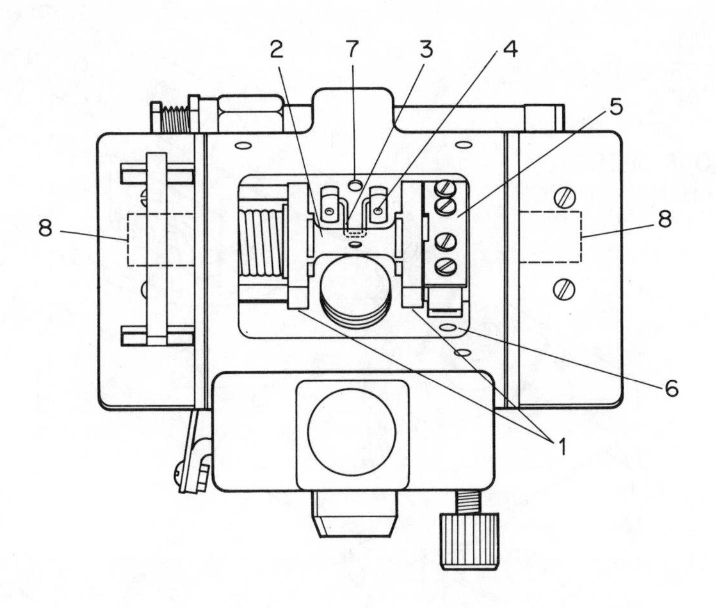

Fig. 5.2c *Graphite furnace assembly – top view*

1. Spring loaded electrodes
2. Graphite tube
3. Temperature sensor (resistance)
4. Temperature sensor screws
5. Optical mask (to prevent radiation from the hot furnace reaching the detector
6. Purge gas inlet
7. Purge gas exit
8. Flow deflectors (to prevent fogging of cell window during drying or ashing

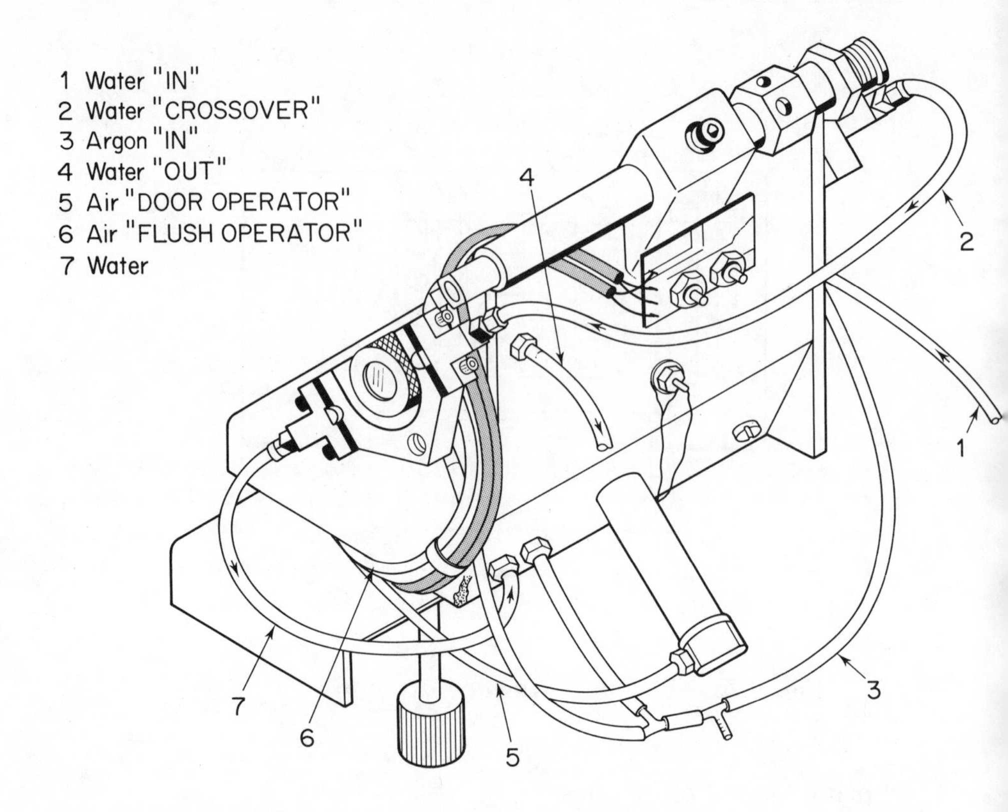

Fig. 5.2d. *Graphite furnace assembly – rear view*

The graphite tube described above is not the only design available. Two different electrothermal atomisers are shown in Fig. 5.2e:

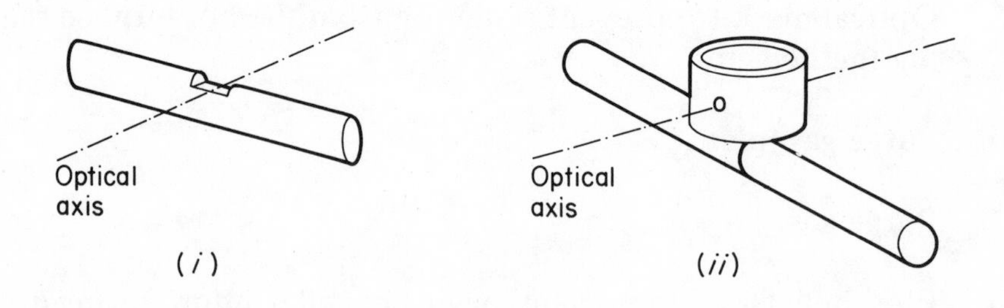

Fig. 5.2e. *Alternative electrothermal atomisers*
(i) the heated graphite filament; (ii) the heated graphite cup

In both designs the atomiser is shielded from oxygen in the air by flowing argon or nitrogen. The various designs will give different atomic residence times. Interferences can also arise in some designs if the atom cloud cools too rapidly.

Π Which design, the filament, the cup or the tube, will give:

(*i*) the greater sensitivity;

(*ii*) the least interference due to rapid cooling of the atom cloud?

(*i*) The atom residence time will be much longer in the tube design where the atoms are confined for about 0.5 – 1.0 s, than the cup or the filament where the atom cloud which will be expanding rapidly due to the large temperature increase, can rise unhindered above the light beam.

(*ii*) The tube furnace will also keep the atom cloud at a higher temperature for longer than the other designs, and so will suffer least from interferences due to rapid cooling.

The tube furnace is the most widely used design, as it gives greater sensitivity, and does not give such strong temperature gradients in the atom cell.

5.3. PROBLEMS IN ELECTROTHERMAL ATOMISATION

We have seen that the use of electrothermal atomisation in furnace-AAS gives a marked improvement in sensitivity over flame-AAS. The furnace technique is not without its own problems, which we shall consider in this Section. In some cases the problems are relatively easily overcome, while in other cases the problems remain a fundamental limitation for furnace-AAS.

5.3.1. Background absorption

Although there are no flame gases in furnace-AAS to absorb ultraviolet light, intense background absorption, often at the 90% level, is observed. This is a fundamental problem in furnace-AAS, as it is due to the high residence time of the sample in the observation region – just as the atomic absorption is increased so is the molecular absorption. An especially difficult case arises in the direct analysis of solids where a great deal of particulate material may be generated during atomisation to form a 'smoke' which reduces light transmission through the furnace. The effect is the same as increased molecular absorption. There are two ways in which the problem may be overcome – furnace programming and background correction, both of which are essential for reliable use of furnace-AAS.

The aim of *furnace programming* is to get rid of as much light-absorbing material as possible before the atomisation is carried out. Three stages are normally used.

Drying at about 100 °C to remove solvent, taking care to avoid loss of sample by spitting.

Ashing at about 400–500 °C to pyrolyse organic matter. At these temperatures most organic molecules break down to small volatile molecules which are flushed from the furnace.

Atomisation at 2000–3000 °C to give a rapid absorption peak.

A typical furnace programme, represented as a temperature-time profile, is shown in Fig. 5.3a.

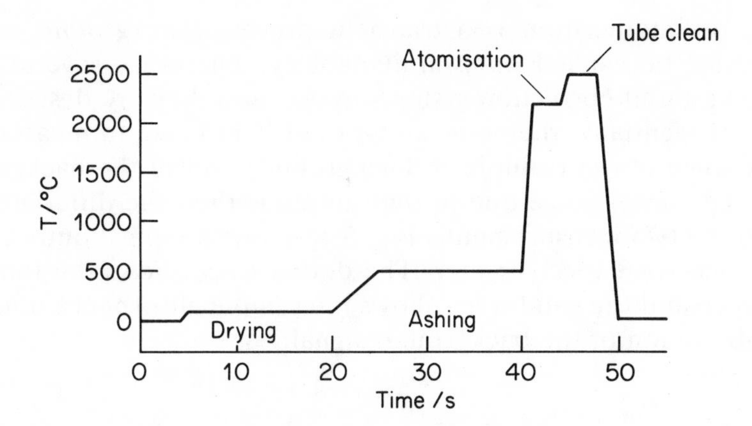

Fig. 5.3a. *A furnace programme*

Fig. 5.3b shows the type of transient absorption signal which is obtained. Either peak heights or peak areas may be used for measurement.

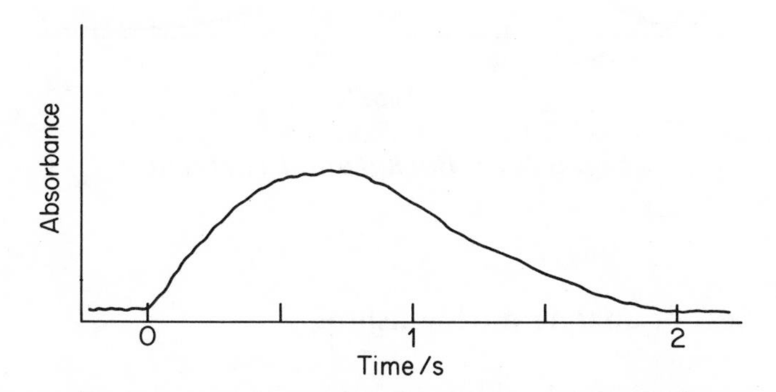

Fig. 5.3b. *Transient absorption during electrothermal atomisation*

The atomisation temperature used is chosen to be just high enough to give rapid and complete atomisation, since excessive use of very high temperatures shortens the life of the furnace.

Since the atomisation is a transient process, *background correction* must be carried out simultaneously. There are several types of background correction systems in use, and they are described in Part 6. Essentially, they use two types of light beam to measure the absorbance of the (sample + background) and of the background only. The absorbance due to the sample is then the difference between the two measurements. Fig. 5.3c shows a typical output from a furnace-AAS spectrometer. The dotted trace shows the total absorbance and the solid trace shows the sample absorbance obtained by subtraction of the background signal.

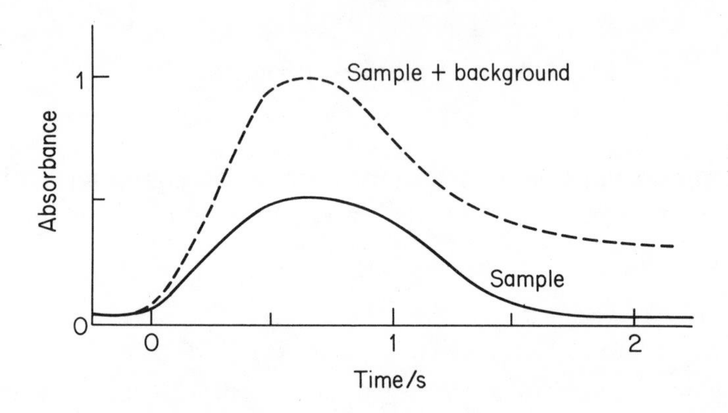

Fig. 5.3c. *Background correction*

5.3.2. Loss of analyte during ashing

Some salts, particularly chlorides and salts of cadmium, arsenic and mercury, are volatile and can be lost during the ashing stage. Lower ashing temperatures should be avoided as this can increase the background signal, but there are various ways to stabilise the elements by forming less volatile compounds eg.

Chlorides can be converted to nitrates by eg adding conc. HNO_3 and boiling off the HCl.

Arsenic can be stabilised by addition of nickel ions to form involatile nickel arsenide.

Mercury can be stabilised by addition of sulphide ions to form mercury sulphide.

The relative volatility of chlorides is a disadvantage in furnace-AAS where they can escape before complete atomisation, although by contrast, their volatility is an advantage in flame-AAS where dissociation to atoms in the flame is speeded up.

The best samples for furnace-AAS analysis are salts such as oxides and oxy-acids, which are so involatile that the atomisation process takes place on or near the surface of the furnace. Reactions of the oxide with the graphite furnace may occur.

$$MO\,(s) + C(s) \rightleftarrows M(g) + CO(g)$$

The equilibrium lies well to the right at the high temperatures involved in the atomisation step.

5.3.3. Memory effects

Sometimes not all of the sample is volatilised during atomisation. This is usually overcome by cleaning the furnace between samples by firing at higher temperatures than used for atomisation.

The memory effects due to incomplete volatilisation are partly caused by the highly porous nature of graphite, which allows the sample to diffuse into the furnace walls. The effects are reduced with the use of *pyrolytic graphite* furnaces, formed by depositing carbon onto the surface of the furnace, from an electrical discharge in methane, to reduce the porosity. Pyrolytic graphite also has the advantage of reducing metal carbide formation, which can slow the rate of atomisation.

5.3.4. Temperature gradients

The ideal furnace would be *isothermal*, that is, the wall and gas would all be at the same, constant temperature. In practice the gas inside the furnace heats up more slowly than the furnace wall. If the atoms are formed rapidly on the wall of the furnace and migrate rapidly into a relatively cool gas, *condensation* can occur if the gas is too cool to maintain atomisation. This problem was particularly severe with some early designs of electrothermal atomiser which used a filament rather than a hollow tube furnace, since the atoms rapidly cool down as they move away from the filament. Fig. 5.3d shows how the temperature of the sample, T_S, lags behind the temperature of the furnace wall, T_W, once the atoms have escaped into the cooler gas.

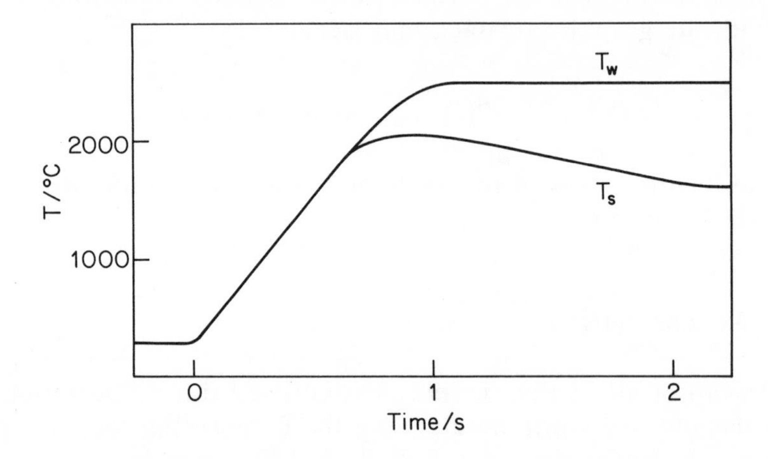

Fig. 5.3d. *Thermal history of furnace sample*

One way to avoid the problem of condensation in furnace-AAS is to put the sample onto a graphite *platform* (sometimes referred to as the L'vov platform) inside the furnace. The platform is not heated directly by the electrical current but indirectly by radiation and convection from the furnace walls. The effect is to delay atomisation until the gas has heated up sufficiently to prevent condensation.

An accessory which has the advantage of delaying atomisation until conditions are closer to isothermal is the graphite boat. The sample is placed on the boat which is then pushed into the furnace. The graphite boat is useful for solids which can be weighed out into the boat.

An alternative to the platform is to use *probe atomisation*. As shown in Fig. 5.3e, the sample is placed on on an external graphite probe which is only inserted into the furnace after the tube has reached a constant temperature. Atomisation at 2500 °C requires about 10 s to establish the constant tube temperature, and then the probe is inserted for about 3 s to atomise the sample.

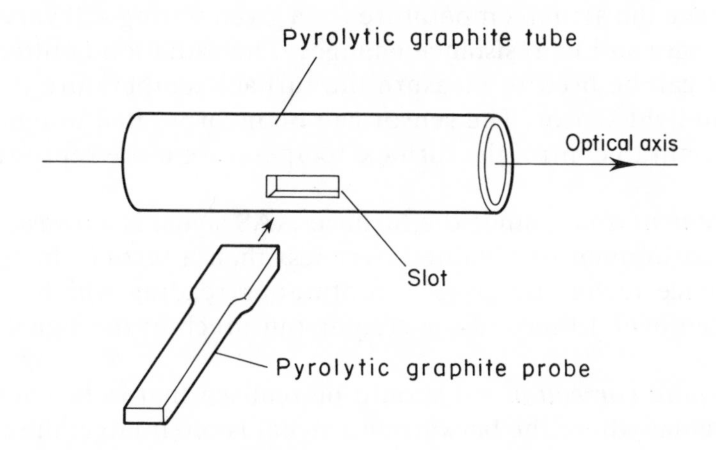

Fig. 5.3e. *Probe atomisation*

An attractive feature of the probe atomisation technique is that it can readily be automated.

5.3.5. Precision

One of the main disadvantages of furnace-AAS is the poor precision which is obtained. With manual injection of the sample into the furnace a precision of 5% is usual. It is not so easy to obtain results with the precision of 1% or so, which is readily obtained for flame-AAS. The poor precision of the furnace technique is due to several factors:

Small sample volumes – errors in pipetting small volumes of up to 0.1 cm^3 are often about 3 to 5% . The reproducibilty can be greatly improved to about 1% by using automatic samplers. In the case of solid samples heterogeneity of the sample can be a problem, since such small volumes are atomised – different portions of sample may contain different amounts of the element.

Reproducibility of furnace temperature – the temperature of the graphite furnace is determined, for a given voltage, by its electrical resistance. If the furnace programmer simply controls the voltage to the furnace the actual temperature for a given setting will vary as the furnace ages and its resistance changes. The radiation emitted from the wall can be used to measure the furnace temperature using an infra-red light sensor. The sensor can be incorporated into an electrical circuit to control the furnace temperature more reproducibly.

Measurement time – since the furnace-AAS signal is a transient, the peak measurement is obtained over less than a second. In contrast the furnace technique gives a continuous reading which is often integrated over 10 seconds, averaging out much of the signal noise.

Background correction – it should be realised that in furnace-AAS in particular where the background signal is often large, the sample signal is obtained as the difference between two large values, the relative error in the difference signal can be much larger than in the two measured signals.

Sample throughput – the complex furnace programme required for drying, ashing and atomisation can take about 2 minutes – roughly 10 times as long as flame-AAS measurements. Averaging a series of readings on a sample to improve precision may be impractical.Use of an automatic sampler can overcome this limitation to some extent by allowing long periods of unattended operation.

Π If the absorbance of the (sample + background) beam is 0.80 ± 0.01 and the absorbance of the background beam is 0.70 ± 0.01, what is the absorbance of the sample? What is the percentage error in the sample absorbance? Assume for simplicity that the errors are additive.

The sample absorbance is 0.80 − 0.70 = 0.10. The error is ±0.02 or 20%. Since concentration is proportional to absorbance, the error in concentration would also be of this size. (A more realistic error treatment would give ±14%).

Note that the percentage error in the two original measurements is only about ±1 or 2%.

Π One way of increasing the precision is to average several measurements to reduce random errors. The term signal to noise ratio is often used as a measure of the reliability of a reading. How many measurements would you need to average to improve the signal to noise ratio by a factor of 10? (The signal to noise ratio is proportional to $n^{1/2}$, where n is the number of measurements.)

If each measurement takes 2 minutes, because of the complex furnace programming needed, how long would it take you to improve your signal to noise ratio by a factor of 10?

The number of measurements needed is 10^2 or 100. The total time needed is 100 × 2 = 200 minutes, or 3 hours and 20 minutes (excluding tea-breaks!). This is a minimum analysis time – you would also have to check standards regularly to allow for instrumental drift. Clearly producing large improvements in precision by averaging results is not realistic.

Poor precision due to sample pipetting errors is usually the main problem when manual sample injection is used, but with the increasing use of automatic analysers, *precision* is no longer considered to be as much a limitation for furnace-AAS as *accuracy*. *Random* errors may be minimised but the technique still suffers from *systematic* or one-way errors. For example incorrect values of concentration can be measured with background correction if the deuterium and hollow cathode lamps are not accurately aligned along exactly the same optical axis.

To summarise, in spite of the advantages of furnace-AAS over flame-AAS, the poor precision, low sample throughput, and background correction problems outweigh the advantages for many routine problems. As a general rule if a sample can be analysed by flame-AAS then it is preferable to do so. On the other hand there are many instances where furnace-AAS is the only satisfactory technique, particularly where very high sensitivity is required or only very small samples are available.

The techniques of flame and furnace-AAS are in fact complementary to each other, and most laboratories use both techniques. Only one spectrophotometer is required, and it is a simply a matter of minutes to remove the burner for the flame and replace it with the furnace or vice-versa.

SAQ 5.3a

Would the peak absorbance shown in Fig. 5.3b be increased or decreased by carrying out the atomisation at a lower temperature?

SAQ 5.3b

In graphite furnace-AAS the large gas expansion which occurs as the sample volatilises, limits the residence time of the atoms in the atom cell. Some manufacturers use a pressurised cell to keep the atoms in the furnace to increase the residence time and hence sensitivity. This can be done by putting windows at the ends of the furnace, or by flowing gas into the cell from both ends. Will background absorption be more or less of a problem than in the unpressurised cell?

SAQ 5.3c

Cadmium in water effluent was analysed by the method of standard additions using furnace-AAS. In one experiment a normal furnace was used and the following results obtained:

	Absorbance
sample	0.078
sample + 0.3 ppm Cd	0.140
sample + 0.6 ppm Cd	0.205

It was felt there might be some suppression of signal due to atomisation off the wall occurring before the gas had heated up so the experiment was then repeated using a platform furnace. The following results were obtained:

	Absorbance
sample	0.090
sample + 0.3 ppm Cd	0.160
sample + 0.6 ppm Cd	0.240

For each set of results calculate the concentration of cadmium in the effluent. Can you say whether the problem of condensation is likely to be present in the normal furnace?

(The method of standard additions was described in Part 4 of the Unit – you should plot absorbance on the y-axis against added concentration of cadmium on the x-axis, and extrapolate back to the x-intercept to obtain the cadmium concentration of the sample. If in doubt check through SAQ 4.1d.)

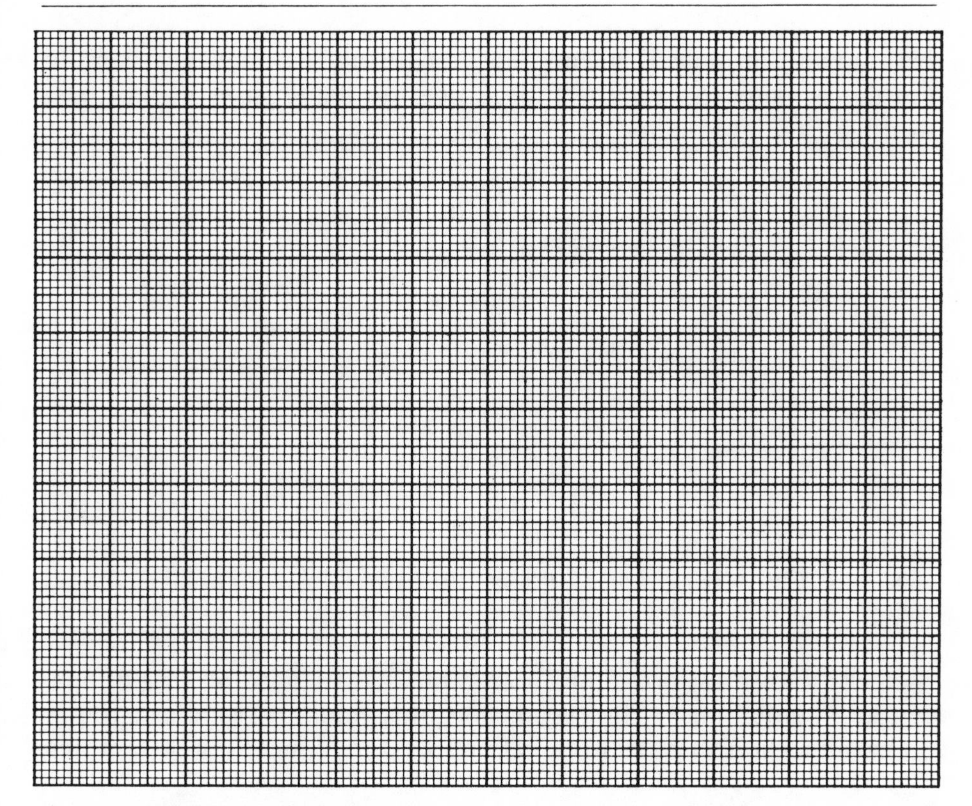

5.4. HYDRIDE METHODS

We saw in Part 5.1 that many elements are difficult to analyse by flame-AAS because of their short wavelength primary resonance lines near 200 nm where absorption by the flame gases is strong. Electrothermal atomisation with background correction can be used to study these elements, but an alternative approach is to generate a vapour of the hydride of the element. The hydride can then be decomposed to an atomic vapour at moderate temperatures, without direct injection into a flame.

The basic system is shown in Fig. 5.4a. The burner assembly in a standard atomic absorption spectrometer is replaced with a silica tube which can be heated electrically by a coiled wire. The hydride vapour is flushed by a stream of argon into the tube where the temperature is sufficiently high (about 900 °C) to atomise the vapour. The atoms are detected by the beam from the resonance lamp which passes down the axis of the tube.

Although the atom cell is flameless, there is no reason why we should not use a flame to heat the tube, since the flame gases do not pass through the light beam.

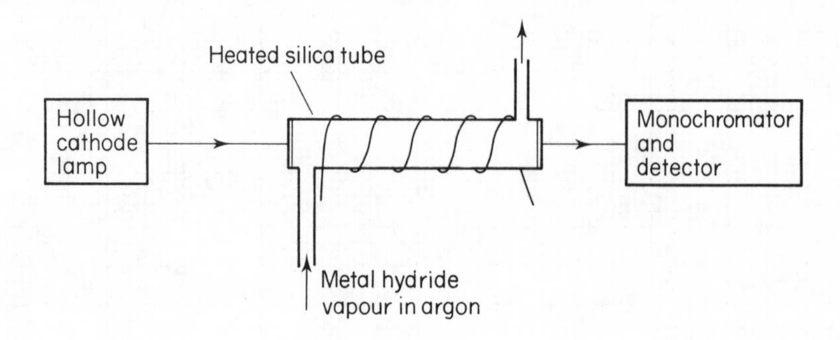

Fig. 5.4a. *The atom cell in the hydride method*

Detection limits approaching 10 ng dm^{-3} (0.01 ppb) or 1 ng of analyte can be obtained. The response is only linear up to about 100 ng so smaller aliquots need to be taken for more concentrated samples.

Π Which type of silica tube will give the best sensitivity?

(*i*) A short or a long tube, or

(*ii*) A narrow or a wide tube.

(*i*) A longer tube will give a greater sensitivity, since there will be more atoms in the light path.

(*ii*) A narrow tube will give a greater sensitivity, since in a wide tube many of the atoms will be outside the region through which the light beam passes.

Fairly simple and cheap equipment is needed to modify an existing spectrometer, and detection limits comparable to those obtained with furnace-AAS are obtained with suitable elements. We shall look now at which elements are suitable for the technique, how the hydrides are generated, and the relative merits of the technique.

5.4.1. Suitable elements

The first requirement of the technique is that the element of interest should form a volatile hydride.

Π Which of the following elements do you think will form volatile hydrides ?

(*i*) Sodium.

(*ii*) Selenium.

Selenium forms a volatile hydride, but sodium does not. This is because elements on the left of the periodic table such as the alkali metals and the alkaline earth metals have a strong tendency to form ionic hydrides which are crystalline and involatile. Sodium hydride is formed as crystalline Na^+H^- (a solid which is stable up to 800 °C). The elements in the middle and on the right of the periodic table are more likely to share electrons with the hydrogen atoms in covalent bond formation. The forces between covalently bonded molecules are very much weaker than the electrostatic forces in ionic crystals, and covalently bonded hydrides are very volatile. Obvious examples are methane, ammonia, and hydrogen sulphide. Since selenium is in the same group of the periodic table as sulphur (group VI) it will form the volatile hydride H_2Se (boiling point 41.5 °C).

The elements most widely studied by the hydride method are in groups IV, V and VI of the periodic table. Examples are germanium (265.1 nm), arsenic (193.7 nm), antimony (217.6 nm) and selenium (196.1 nm). The low wavelength of the primary resonance lines for arsenic and selenium makes them particularly difficult to study by flame-AAS.

5.4.2. Hydride formation

To generate the hydride, the sample is usually added to a solution of HCl (0.5 to 5.0 mol dm^{-3}) and $NaBH_4$ (about 1%) for reaction times of 10 to 100 seconds. The hydride vapour is flushed into the silica tube atom cell at argon flow rates of 20 to 100 $cm^3\ s^{-1}$.

The are two main problems with the hydride formation step:

(*i*) the efficiency of hydride formation depends on the oxidation state of the metal. The problem can be overcome by converting the element to a single oxidation state before the hydride generation step. Group V elements such as As and Sb tend to form both +3 and +5 oxidation states, of which the +3 states form the hydrides more easily, so the elements should be reduced to the +3 state before hydride generation.

(*ii*) the presence of some ions which are easily reduced, such as cobalt, copper and nickel can suppress the hydride formation. This appears to be caused by the reduction of these ions by the sodium borohydride to a finely dispersed precipitate of the free metal. The surface of these metal precipitates can catalyse the decomposition of the hydride. You may remember from Part 4 of the Unit that most elements will dissolve in strong acids, and the problem can be largely avoided by using stronger HCl (5.0 mol dm^{-3}).

5.4.3. Relative merits

The main advantages of the hydride method are the lack of absorption by flame gases, not having to use a nebuliser, and the long residence time of the atoms in the atom cell, which is of the order of seconds. All these factors contribute to the greatly increased sensitivity compared to flame-AAS. By contrast with furnace-AAS, not all of the original sample enters the atom cell, so background correction problems are not so severe, and complicated furnace programming is not required. The equipment for hydride generation is much cheaper than that required for furnace-AAS. The technique is particularly useful for those difficult elements which have low wavelength primary resonance lines, since many of these elements form volatile hydrides.

The main disadvantages of the hydride method are its lack of generality, since only a limited number of elements readily form volatile hydrides, and the problems of inter-element interference in the hydride generation step. The use of a wet method to prepare the hydride is often less attractive when direct determinations can be carried out by furnace-AAS, often without pre-treatment.

SAQ 5.4a

Your laboratory has an atomic absorption spectrometer with facilities for flame, furnace and hydride techniques. Decide which technique you would use for the following measurements.

(*i*) The owner of a fishery is having problems which might be caused by barium leaching from a nearby disused barytes mine into the river upstream of the fishery. Levels of a few mg dm^{-3} of barium are likely to be detrimental to fish.

(*ii*) The waste effluent from a factory which is thought to contain arsenic at a concentration of about 0.1 mg dm^{-3}.

(*iii*) A forensic scientist gives you a few milligrams of plastic for identification. The type of plastic is known but the manufacturer of the plastic needs to be established. One manufacturer uses zinc oxide as a flame retardent, while another uses antimony oxide as the flame retardent. Both oxides would be present at levels of a few percent (w/w) of the plastic.

SAQ 5.4a

5.5. COLD VAPOUR DETERMINATION OF MERCURY

Mercury is something of a special case since it is the only element, apart from the inert gases, which has a high vapour pressure of atoms at room temperature (the saturation vapour pressure is about 0.1 Pa at 20 °C). Because of this we can use a procedure for mercury determination which is similar in principle to the hydride method. A reduction step with either sodium borohydride or tin (II) chloride, is used to reduce inorganic mercury compounds to elemental mercury which evolves from the solution :

$$Hg^{2+} + Sn^{2+} \rightleftharpoons Hg + Sn^{4+}$$

The Hg vapour is flushed out, by bubbling argon through the solution, along a pyrex or silica tube, for analysis using a hollow cathode mercury lamp. The optical arrangement is shown in Fig. 5.5a and is similar to that described in Fig. 5.4a for the hydride method. The main difference is that there is no need to heat the silica tube since the mercury is already in atomic form. The design in Fig. 5.5a also differs in that the cell is open-ended. This allows the use of the most sensitive 184.9 nm resonance line without reduction of the light intensity by absorption at the cell windows.

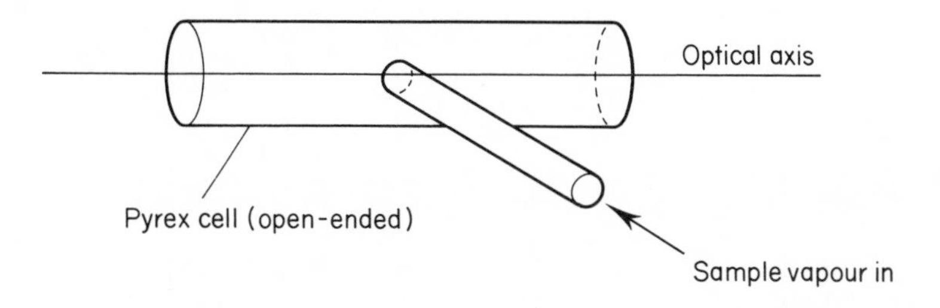

Fig. 5.5a. *A mercury vapour cell*

For flame-AAS the detection limit is rather high, about 0.2 ppm, because the 184.9 nm resonance line cannot be used due to the flame gas absorption, so that the much less sensitive 253.7 nm line has to be used. With the use of a flameless atom cell the more sensitive line can be used. The longer residence time in the atom cell also contributes to the improved detection limit (a factor of about 100) compared with flame-AAS.

Organic mercury compounds, which are of great environmental significance, are oxidised to inorganic mercury before analysis.

The method has been widely used, but background correction is usually needed and some very reducible elements, particularly the noble metals such as gold, can interfere with the reduction.

SAQ 5.5a

0.1 cm^3 of an aqueous solution containing 0.01 mg dm^{-3} of mercury was analysed by the cold vapour method. Can you work out how many atoms of mercury (A_r (Hg) = 200.6) are present?

5.6. OTHER METHODS

Although technically the three techniques described in this Section do not use a flameless atom cell, they are something of a cross between flame atom cells and the flameless techniques described in this Part of the Unit. All the techniques use a tube mounted in the flame to act as an atom trap, increasing the residence time of the atoms in the atom cell and improving sensitivity over normal flame methods.

(*i*) The *Delves cup* has in the past been widely used for the analysis of microsamples of blood for lead. The important features of the Delves cup are illustrated in Fig. 5.6a. The sample is placed on a small nickel or tantalum boat which is rapidly inserted into the flame just below a tube of silica, ceramic or nickel. The sample is vaporised and atomised and flows into the tube which serves to keep the atoms in the optical path for much longer than in a conventional flame.

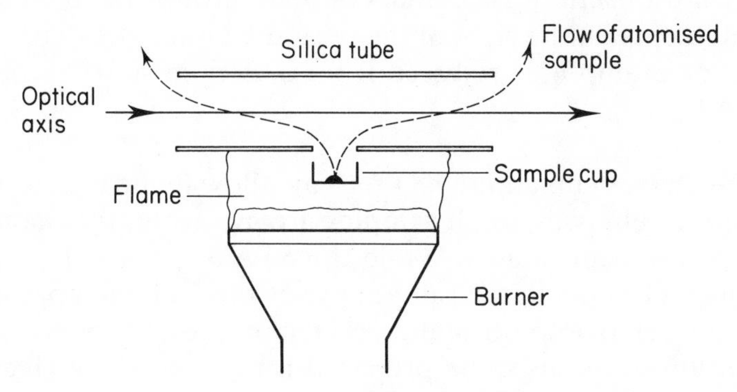

Fig. 5.6a. *The Delves cup modification to flame-AAS*

(*ii*) A development of this technique is the *slotted tube atom trap*, or STAT, which simply involves positioning a silica tube directly above the flame with a narrow slot at the bottom and another at the top of the tube as shown in Fig. 5.6b. Some of the atoms in the flame enter the bottom slot and are trapped in the tube for some time before flowing out of the exit slot at the top.

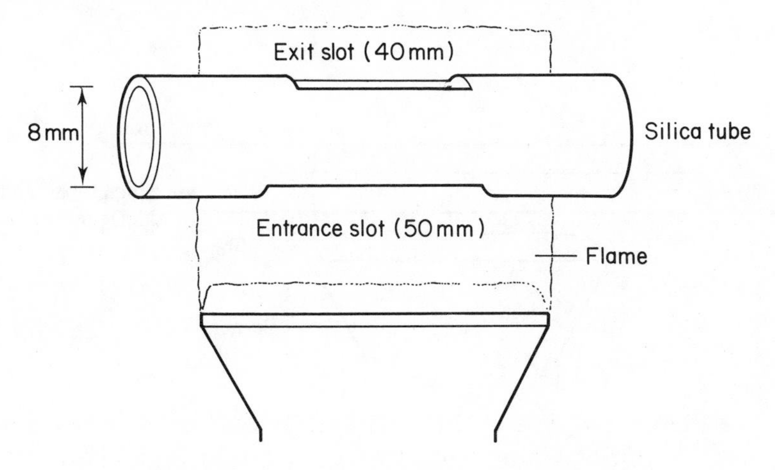

Fig. 5.6b. *The slotted tube atom trap*

The technique gives improvements in detection limits for some elements by a factor of 2 to 10 compared to normal flame-AAS. The technique is limited to use with those elements which can be analysed with an air–acetylene (or air–hydrogen) flame since silica cannot withstand the temperatures of a nitrous oxide flame. The technique is most useful for elements of main groups IV to VI and transition elements on the right of the periodic table. The detection limit for tin, for example, is reduced from 0.06 ppm to 0.015 ppm using the STAT.

Both the Delves cup and the STAT can allow flame techniques to be used where relatively small samples are available, for example with blood serum from infants. While the STAT is not strictly a microsampling technique in the same sense as the Delves cup, the greater sensitivity, relative to normal flame techniques, of the STAT allows more dilute solutions to be prepared for analysis to a given degree of precision. On the other hand the sensitivity improvement is not so dramatic as for the flameless atom cells described earlier in this Part of the Unit.

(*iii*) An alternative approach to increasing the residence time of the atoms in the flame is to *pre-concentrate* the atoms in the flame by *atom-trapping atomic absorption spectrophotometry* (ATAAS). A water-cooled silica tube is suspended in the flame as shown in Fig. 5.6c:

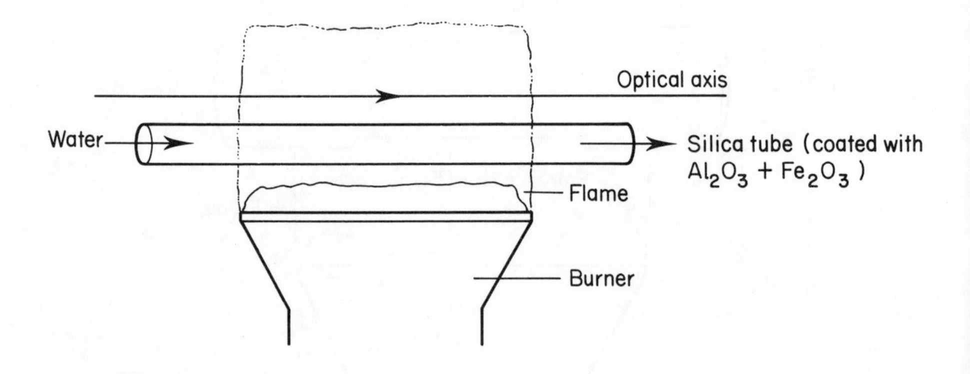

Fig. 5.6c. *Atom-trapping-AAS*

The surface of the silica tube is cooled to 350 to 400 °C by the flowing water. This is cold enough for the atoms of some elements such as cadmium to condense as the metal. (Other elements such as calcium, which form very stable oxides, will condense as the oxide on the tube.)

Collection times of a few minutes are used to allow the atom concentration to build up. Then the tube is rapidly heated, by removing the water coolant with a blast of compressed air. The atoms are liberated as a concentrated cloud, giving a transient absorption signal.

Although more time-consuming than conventional flame-AAS, the ATAAS technique has much lower detection limits. The detection limit for cadmium, for example, can be reduced 30-fold to about 30 to 100 ng dm^{-3} (0.1 ppb).

Π Cadmium in soil may be analysed by extraction with calcium chloride solutions. Unfortunately calcium in flame-AAS can interfere with cadmium analysis. Is the ATAAS technique likely to be prone to interference from calcium to the same extent?

The calcium interference will be reduced in the ATAAS technique, since calcium will condense on the silica tube as the oxide as stated above. The oxide is thermally very stable and will only volatilise very slowly from the silica. The cloud of atoms of cadmium formed when the tube is heated will contain far less calcium than in conventional flame-AAS, and the interference from calcium is greatly reduced.

SUMMARY AND OBJECTIVES

Summary

Several techniques have been developed aimed at overcoming the sensitivity limitations of flame-AAS. The most widely used and widely applicable technique uses electrothermal atomisation to form the atom cell, and is usually referred to as furnace-AAS. In this technique the residence time in the atom cell is greatly increased and the nebuliser and flame are not needed. The main advantages are the much greater sensitivity, often by a factor of 1000, enabling ppb levels to be routinely determined, the easier analysis of elements at short wavelengths, and the ability to use very small samples. The main disadvantages are the poorer precision relative to flame-AAS, the frequent need for high levels of background correction, and the longer analysis times. The techniques of hydride generation and cold vapour determination are also useful for a number of elements, particularly some of those which have short wavelength primary resonance lines and which form volatile hydrides. Various modified flame techniques – the Delves cup, the STAT and the ATAAS – can also give improved detection limits in some cases by improving the atom residence times.

Objectives

You should now be able to:

- describe the limitations of flame atomisation techniques;
- explain how the processes of electrothermal atomisation, using a graphite furnace, can overcome many of these limitations;
- describe the problems which can arise in furnace-AAS, and explain to what extent these problems can be overcome;
- identify which type of flame or flameless atom cell is the most appropriate for a given analysis;
- identify which elements may be usefully analysed by the hydride generation method.
- understand the basic principles of the cold vapour method for mercury determination, and why the Delves cup, STAT and ATAAS methods can give improved sensitivity compared to conventional flame-AAS.

6. Background Correction in Atomic Absorption Spectroscopy

Overview

We shall see in this section that the measured sample absorbance is often higher than the true absorbance. This can be due to absorption by molecules or light scattering by small particles in the atom cell.

There are three widely used techniques for background correction, and we shall consider their relative merits in this section, following a consideration of the principles of the three techniques.

6.1. CAUSES OF BACKGROUND INTERFERENCE

It is important to realise that we are not concerned here with background absorption which is continually present, such as that due to species such as CH, C_2 or OH in a flame spectrometer, since these are allowed for when the blank solution is analysed. We are instead concerned with changes in absorption occurring when the sample is introduced into the atom cell which are not caused by the analyte atoms, but by the 'matrix'. By matrix we mean all the species other than the one we are interested in the sample. For example the presence of high salt concentrations in a solution being analysed can cause light scattering as an aerosol is produced in the atom cell.

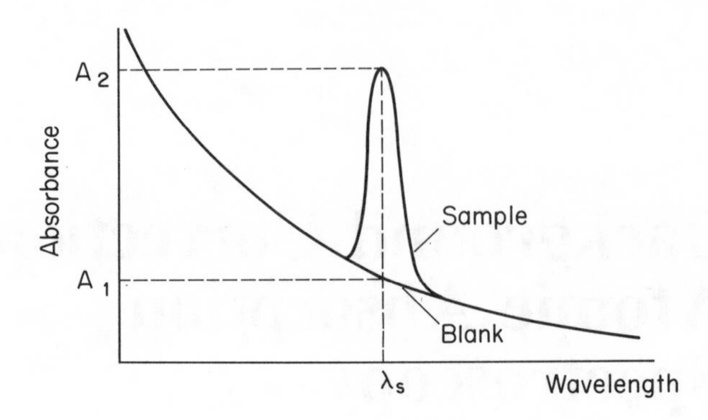

Fig. 6.1a. *Absorption spectrum of sample and blank*

First let's look at how the absorption spectrum might look when these background interferences are not present. Fig. 6.1a shows a plot of absorbance versus wavelength in the region of interest. The wavelength used for analysis is λ_S. Note that in addition to the sharp atomic absorption line there is a broad background spectrum due to molecular absorption and light scattering. However this background is also present in the absence of the sample so that we can measure the true absorbance of the sample, A_S simply by subtracting the absorbances with and without the sample, A_2 and A_1, that is:

$$A_S = A_2 - A_1$$

In practice of course we adjust the value of A_1 to zero so that the readout gives the true sample absorbance directly.

Now let's look at a case where background interferences are present. Fig. 6.1b should be compared with Fig. 6.1a. The main point to notice is that the background absorption is different when the sample is introduced, due to the presence of non-specific (background) absorption. Setting A_1 to zero as before and measuring A_2 will not give a true sample absorbance in this case.

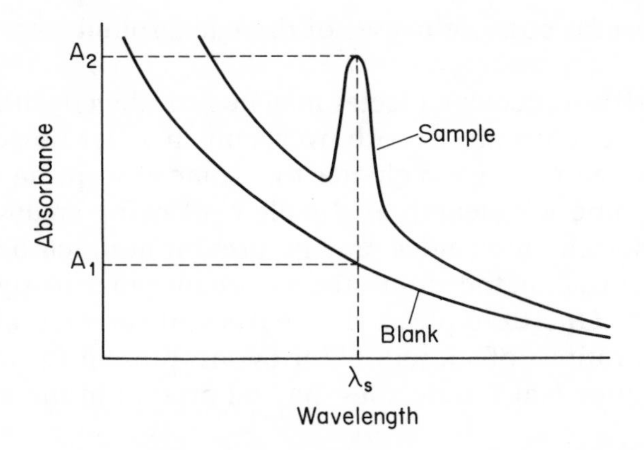

Fig. 6.1b. *Absorption spectrum of sample and blank in the presence of background interference*

Π Will setting A_1 to zero lead to an overestimate or an underestimate of concentration?

The absorbance due to the sample will be overestimated. Since absorbance is proportional to concentration, the sample concentration will be overestimated. As Fig. 6.1c shows the sample absorbance, $A_2 - A_3$, is lower than the measured absorbance, $A_2 - A_1$.

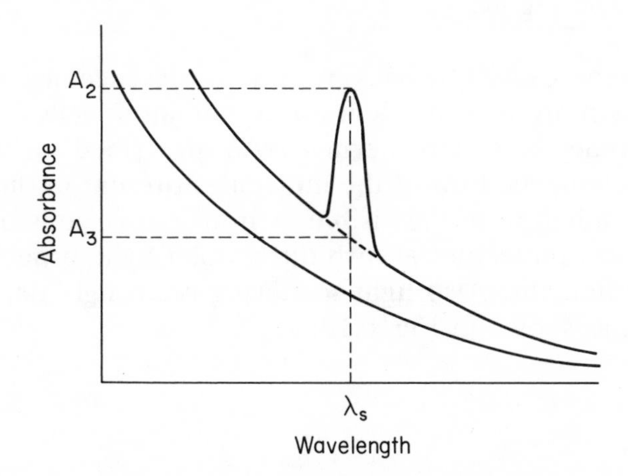

Fig. 6.1c. *Comparison of sample absorbance and measured absorbance in the presence of background interference*

There are several possible causes of the background interference.

(*i*) Absorbing molecular species may be introduced with the sample. Since molecules absorb over a much wider range of wavelengths, there is a good chance that some absorption will occur at the same wavelength as the element being analysed. Since the absorbing molecules may be present at much higher concentrations than the atoms, the absorbance may be significantly affected. One example is the analysis of samples where high concentrations of sodium chloride are present in the sample, as diatomic NaCl molecules may be present in the atom cell.

(*ii*) More rarely, absorbing atomic species may also be present in the sample. For example, if antimony is present in a sample of nickel, the 231.147 nm antimony line may interfere with the determination of nickel at 231.096 nm.

(*iii*) The introduction of particles into the atom cell can cause light scattering. This is important at high salt concentrations and will lead to an apparent increase in absorption since less light reaches the detector. Light scattering is a common, but fortunately not an everyday, experience for most of us in the form of fog or smog. It is the scattering of the sunlight by the small particles in the atmosphere which reduces the amount of sunlight reaching us.

These problems caused by background interferences are not universal. They are more of a problem for certain atom cells (particularly graphite furnace or electrothermal AAS, described in Part 5), and for certain elements. Few of the molecules present in the atom cell can absorb radiation in the visible or infra-red regions of the spectrum, but most molecules absorb ultra-violet light, especially at the shorter wavelengths. Also light scattering is strongly dependent on wavelength, as shown in Fig. 6.1d.

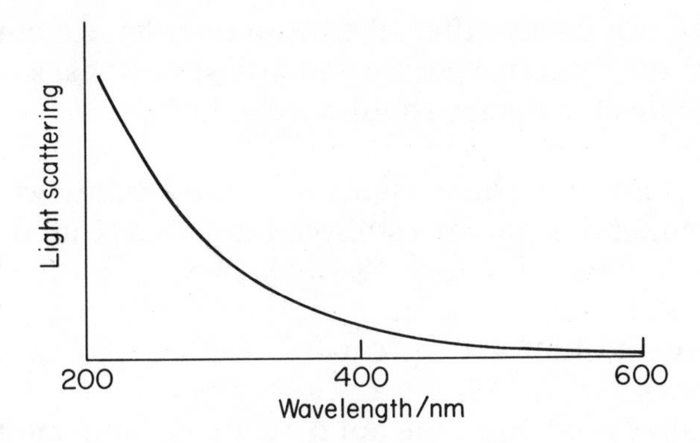

Fig. 6.1d. *Wavelength dependence of light scattering*

Π Which analysis is more likely to involve background interference – barium at 553.5 nm or arsenic at 193.7 nm?

Since light scattering is much weaker and molecular absorption is negligible in the visible region of the spectrum, the barium analysis should present fewer problems than the arsenic analysis.

As a general rule, background correction is usually only needed for analyses in the ultra-violet region of the spectrum, up to wavelengths of around 350 nm.

6.2. PRINCIPLES OF BACKGROUND CORRECTION TECHNIQUES

The three commonly used techniques are:

(*i*) deuterium arc background correction:

(*ii*) Zeeman effect background correction, and

(*iii*) Smith–Hiefte (pronounced Heef-yeh) background correction.

Although the principles differ, the techniques have a common aim – to measure the total (ie specific and non-specific) absorbance and the non-specific absorbance simultaneously.

We shall first look at the deuterium arc method as this was the first to be introduced, and is perhaps still the most widely used technique.

6.2.1. The deuterium arc system

This system uses two lamps, the hollow cathode lamp emitting a very narrow linewidth of typically around 0.002 nm, and a deuterium arc lamp, emitting radiation over a very wide wavelength range. The deuterium arc lamp is simply an electrical discharge between two electrodes in an atmosphere of deuterium gas. Most of the emitted radiation is from electronically excited *molecules* of deuterium, and is therefore over a wide wavelength range, as shown by Fig. 6.2a.

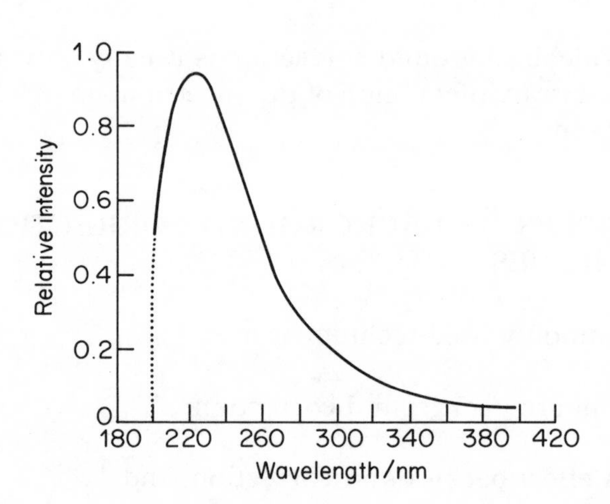

Fig. 6.2a. *Emission spectrum of a deuterium arc lamp. (This is a simplified spectrum in that some sharper atomic emission lines also present are left out for clarity)*

In practice some form of beam splitter such as a rotating sector, which contains alternately transparent and reflecting surfaces, is used to pass the beams through the atom cell in rapid alternation, as shown in Fig. 6.2b. The detector will generate an alternating voltage as the two beams reach the detector.

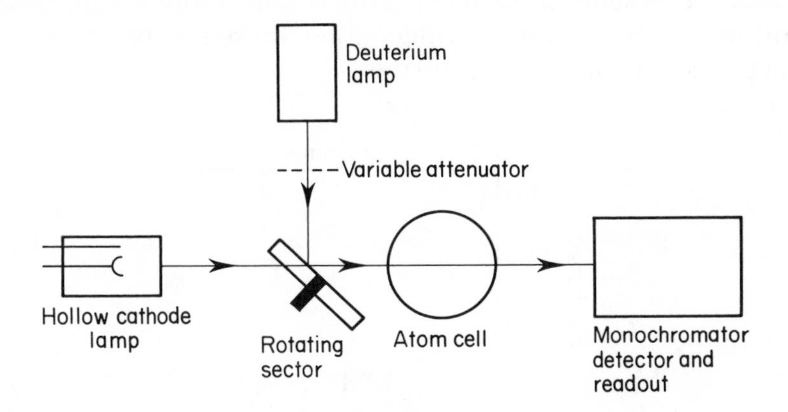

Fig. 6.2b. *Optical system for deuterium background correction*

The spectrometer is set up using the variable attenuator to match the intensities of the two light beams. The difference signal is then zero. When the sample is introduced into the atom cell, the signal reaching the detector is of the form of a square wave, and the alternating voltage generated is proportional to the difference in the absorption of the two beams. The light from the hollow cathode lamp is absorbed by both the sample and the background, but the light from the deuterium lamp is absorbed almost exclusively by the background. The difference between the two beams is then related to the sample absorption only.

The principles underlying deuterium arc correction can take a little time to work out so don't worry if you find the following discussion difficult at the first attempt.

To break the argument up into stages, let us first show that a broad band light source will give only a negligible reading for a very sharp absorption line. Fig. 6.2c shows the light reaching the detector through a monochromator with a spectral bandpass of 0.2 nm.

Even though the peak absorbance for a sharp atomic line of width 0.004 nm is high (0.6), the detector is receiving light at all wavelengths within the monochromator bandpass and hardly notices the amount absorbed by the atom cell. The measured absorbance in this case would be about 0.012. (ie 0.6 × 0.004/0.2 assuming the line shape to be roughly triangular). This is only about 2% of the peak absorbance value. Using a wider monochromator bandpass would reduce this percentage even further.

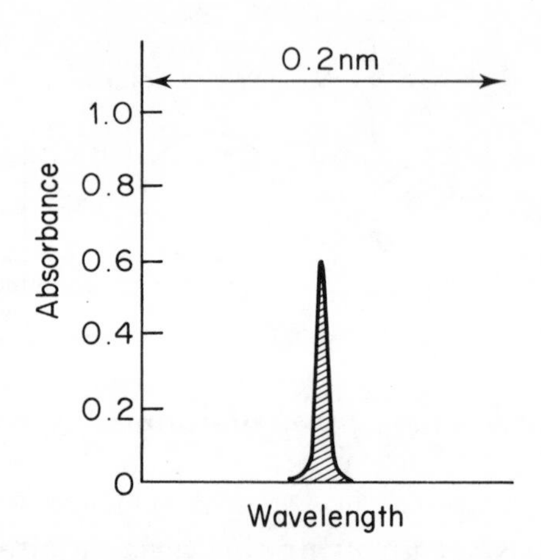

Fig. 6.2c. *Measurement of absorbance using a broad band source (no background signal present)*

Now let us consider what happens when a background signal is present. Fig. 6.2d shows the narrow atomic absorption band superimposed on the flat background signal. (Over the typical range of the monochromator bandpass of 0.1 to 1.0 nm, the spectrum due to the background absorption can usually, but not always, be assumed to be flat). In (*i*), the shaded portion shows the light absorbed from the hollow cathode lamp, while in (*ii*) the shaded portion shows the light absorbed by the deuterium lamp. (Note that the drawings are not exactly to scale.)

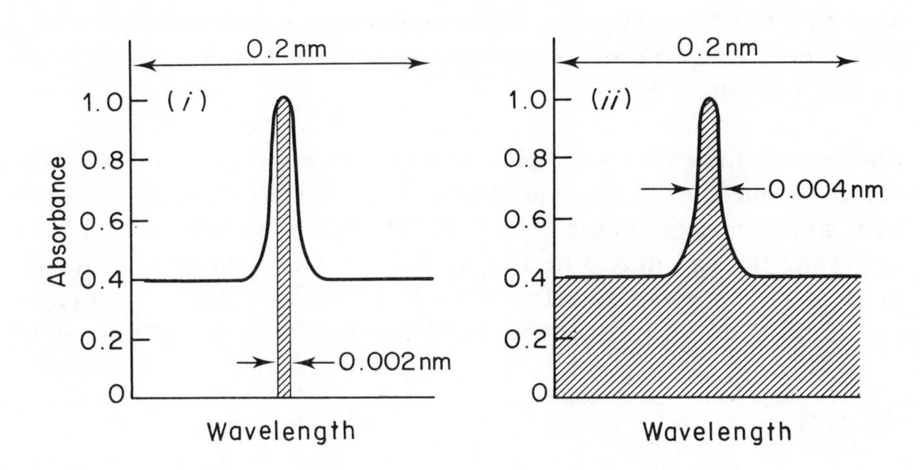

Fig. 6.2d. *Light absorption from (i) hollow cathode lamp and (ii) deuterium lamp. Monochromator bandpass of 0.2 nm assumed*

Π Can you work out from the two diagrams, Figs. 6.2d (*i*) and (*ii*), what the average value of the absorbance will be in each case – A, B, C or D.

Use these values to estimate the absorbance of the sample.

A 0.4 B 0.6 C 1.0 D 1.4

The average absorbance in (*i*) is 1.0 (ie C correct) and for (*ii*) the absorance is 0.4 (ie A correct). The sample absorbance is 0.6.

In (*i*) the lamp emission has a much narrower linewidth than the sample absorption line. Therefore only the centre of the absorption line is sampled and the absorbance is 1.0 .

In (*ii*) the lamp emission is very much broader than the sample absorption, and an averaged value of the absorbance taken over the whole bandpass of the monochromator. Bearing in mind that the absorption peak is actually much narrower than shown the average absorbance will be only slightly above 0.4.

The absorbance due to the sample is then simply (1.0 − 0.4) or 0.6. Note that this is the absorbance we would measure if we were to scan the spectrum at very high resolution or if there were no background signal as in Fig. 6.2c.)

We can see then that the deuterium arc system copes well with the common but simple case described. Let's see whether it can cope with some more difficult problems. Three are shown in Fig. 6.2e. In (*i*) the background is no longer flat, but slopes gradually. In (*ii*) there is overlapping structured *molecular* absorption. In (*iii*) an overlapping *atomic* absorption line from another element is present.

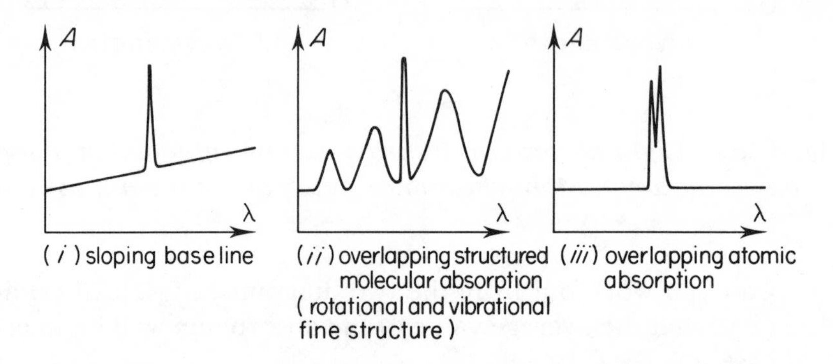

Fig. 6.2e. *Some background correction problems*

Π Do you think that the deuterium arc system will cope with the three situations described in Fig. 6.2e? The important point to remember is whether the *average* value of the background absorbance is the same as the *actual* value of the background absorbance at the wavelength of the sample absorption.

In case (*i*) the background correction will be satisfactory, as long as the atomic absorption line is in the centre of the spectral bandpass of the monochromator.

In case (*ii*) the average background is not necessarily the same as at the atomic absorption line. If the line is on top of an underlying molecular absorption peak, the background will be undercorrected.

An example of this problem can arise in the determination of arsenic at 193.7 nm. If phosphate is present. P_2 molecules in the atom cell can absorb light in this region with relatively sharp rotational and vibrational structure.

In case (*iii*) the background cannot be corrected for if the interfering line is close enough to absorb some of the light emitted by the hollow cathode lamp.

To summarise, the deuterium arc system is a tried and tested, widely used system which copes well with many background correction problems. The main failing of the system is its inability to compensate for sharply varying background signals. Other problems are that the two light beams are often difficult to align exactly along the same light path through the atom cell, and only lines in the ultra-violet can be corrected for. The way round these problems is to use a single light beam and to measure the background absorbance as close as possible to the atomic absorption line. This approach has been used in both the techniques described next.

6.2.2. The Zeeman effect system

This system uses an intense magnetic field to broaden either the emission spectrum of the lamp or the absorption spectrum of the sample. This effect allows the background signal to be measured very close to the atomic absorption line.

The magnet can be placed round either the lamp or the atom cell, as shown in Fig. 6.2f. Both types of instrument are commercially available, although the most commonly used orientation is with the magnet around the atom cell. The theory for both systems is similar and we shall only discuss one system here – the magnet round the lamp. (It should be borne in mind that the hollow cathode lamp design must be modified since magnetic fields affect the operation of the lamp.)

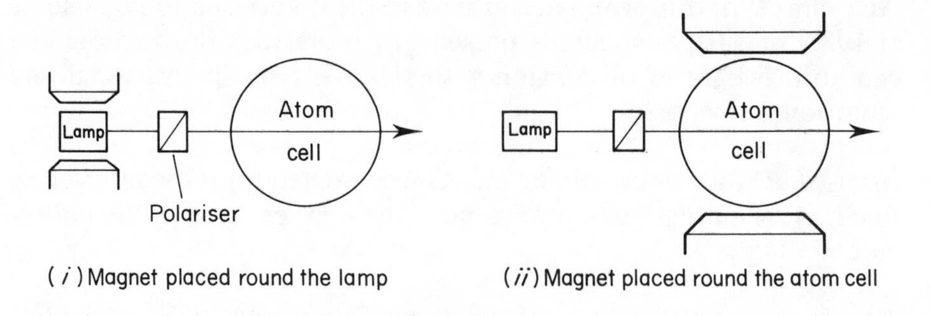

Fig. 6.2f. *Experimental arrangements used for Zeeman effect systems*

An atomic spectral line contains several types of transition (depending on the electron and orbital spins) which normally have exactly the same energy – they are said to be degenerate. In a magnetic field, B, the transitions no longer have exactly the same energy, as shown in Fig. 6.2g. In the example shown, a transition is still observed at the same energy as in the absence of a magnetic field, (the π component), but two new transitions are observed at lower and higher energies (the σ components).

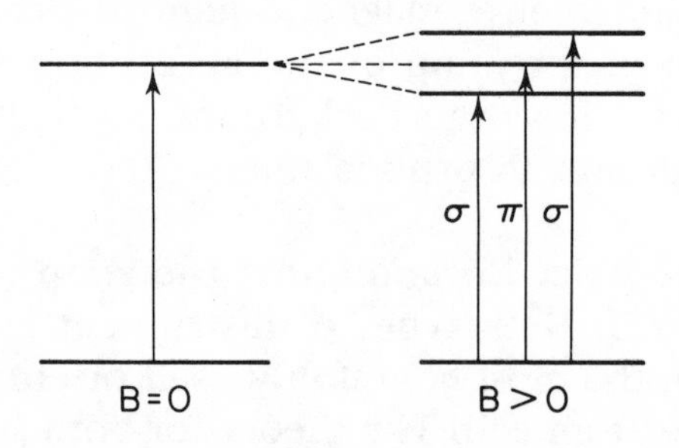

Fig. 6.2g. *The effect of a strong magnetic field on atomic transitions*

Rather powerful electromagnets, producing fields of about 1 Tesla are needed to separate the transitions by more than the spectral linewidth. This is described in Fig. 6.2h. We can use a polariser to isolate either the central, π, line of the spectrum or the wing, σ, lines, since the central line is polarised parallel (||) to the direction of the magnetic field, while the wing lines are polarised perpendicular (⊥) to the direction of the magnetic field. The effect of the polariser is to isolate either the parallel or perpendicularly polarised light, rather like wearing polarised spectacles to remove the polarised light reflected from the surface of the sea. Fig. 6.2h demonstrates how the Zeeman effect is utilised in practice.

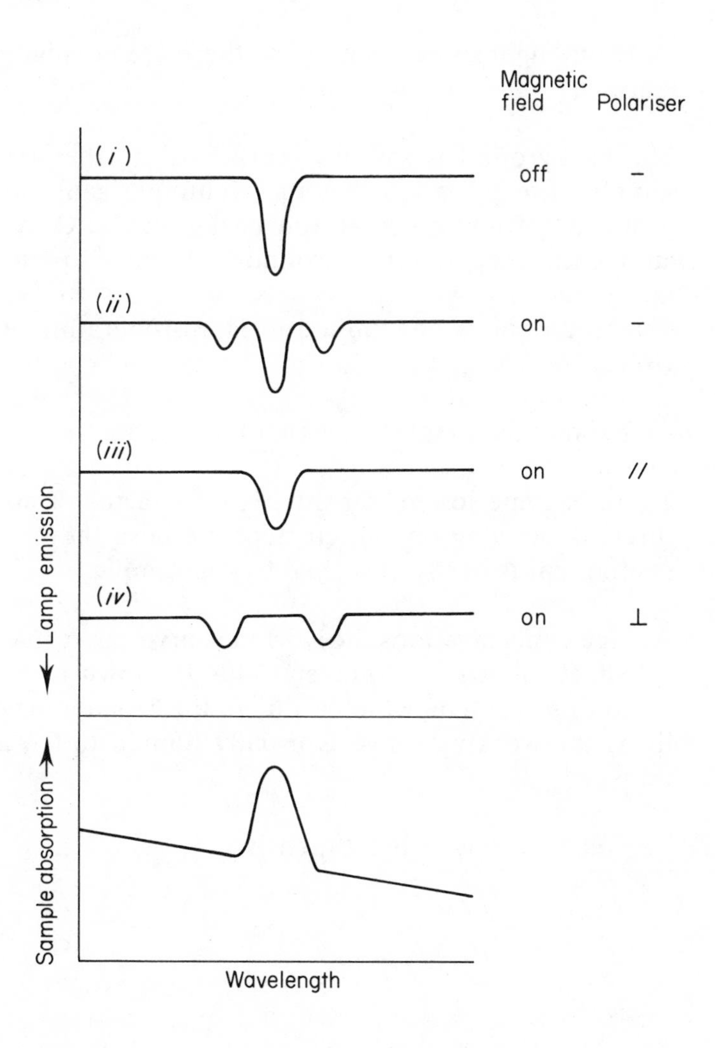

Fig. 6.2h. *Zeeman background correction*

The lamp emission line (*i*) is split into three components by the magnetic field (*ii*). The polariser is then used to isolate the central line (*iii*) which measures the absorbance at the peak of the spectrum, A_π. The polariser is then rotated and the wing lines are isolated (*iv*) and these measure the absorbance of the background A_σ. The sample absorption is then $A_\pi - A_\sigma$.

The most important advantages of the technique are:

(*i*) Only one light source is used, so there are no alignment problems.

(*ii*) The background is sampled very close to the wavelength of sample absorption, so that the technique can cope with the problems of highly structured backgrounds. Only very close atomic lines separated by around 0.01 nm or so are a problem.

(*iii*) The technique is not limited to lamps operating in the ultraviolet.

There are some disadvantages, which include :

(*i*) There is some loss of sensitivity, by a factor of about two or three, depending on the element, because the perpendicular component is partly absorbed by the sample.

(*ii*) At high concentrations the working curve can turn downwards, an effect known as 'rollover'. This is shown in Fig. 6.2i. To avoid this problem, which is due to the broadening of the wing lines, the working range is usually limited to 0.6 absorbance units.

(*iii*) The technique is rather expensive.

(*iv*) The technique cannot be used with flame atom cells if the magnet is placed around the flame, and is effectively limited to use with furnace-AAS. On the other hand, as we have seen above, special hollow cathode lamps are needed if the magnet is placed around the lamp.

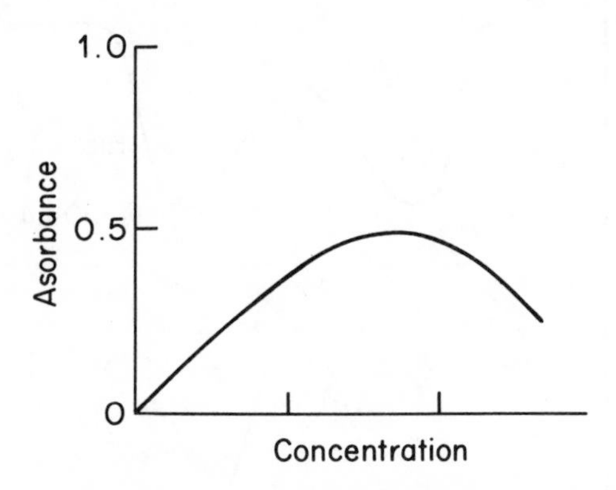

Fig. 6.2i. *Rollover with the Zeeman effect system*

6.2.3. The Smith–Hiefte system

This system is similar to the Zeeman effect system in that a single light source is used, and the spectral output profile of the lamp is modified to measure the background. This is done simply by alternately operating the lamp at low and high current, and is illustrated in Fig. 6.2j. At low current the usual narrow hollow cathode lamp emission line (*i*), is used to measure the absorbance at the peak of the atomic absorption line of the sample. At high current *self-absorption* (described in Part 2) causes a broadening of the lamp output, with a dip appearing in the centre of the emission profile. Thus at low lamp current the total (specific and non-specific) absorbance is measured, while at high lamp current the measured absorbance is essentially that due to the background (non-specific).

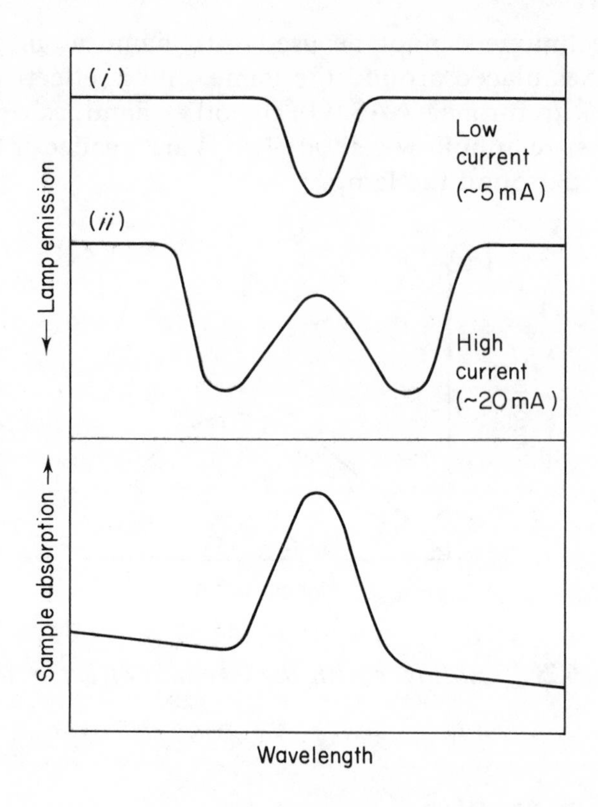

Fig. 6.2j. *Smith–Hiefte background correction*

Π If A_L is the absorbance measured at low lamp current and A_H is the absorbance measured at high lamp current, what will the sample absorbance be?

(*i*) A_H

(*ii*) A_L

(*iii*) $A_L - A_H$

(*iv*) $A_H - A_L$

The correct answer is (*iii*).

Since A_L measures the absorbance of both sample and background, and A_H measures the background absorbance, then the sample absorbance is $A_L - A_H$.

More correctly, there will be some absorption by the sample at high lamp current. This leads, as in the Zeeman system to a loss of sensitivity of a factor of perhaps two or three, depending on the element.

The advantages of the Smith–Hiefte system are essentially the same as those of the more expensive Zeeman systems, namely the use of a single light source of any wavelength, and the ability to correct at a wavelength very close to the sample absorption line. Again highly structured backgrounds can be corrected. No problems due to rollover are found, and the calibration curves are typically linear up to an absorbance of 0.7.

The main disadvantage of the technique is that the electronics cannot be economically incorporated into an existing spectrometer, because of the extensive changes which would be required in the electronics.

Another disadvantage is that when analysing for the more volatile elements the life of the hollow cathode lamp is inevitably shortened by the high current pulses.

An important point to appreciate is that improvements in the analytical method can sometimes reduce the background signal. Where possible efforts should be made to reduce the background signals by appropriate sample pre-treatment (eg furnace programming), since there is a practical limit to how much background absorption can be corrected for. Also the use of background correction methods will reduce the precision of the determination, since it is a difference measurement.

SUMMARY AND OBJECTIVES

Summary

Background interference is usually caused by molecular absorption or particulate light scattering. It is particularly important when working in the ultraviolet with furnace-AAS.

Three techniques are available for the correction of background absorption in AAS.

The deuterium arc correction system uses a second lamp, a deuterium arc lamp, to measure the background signal. This is the most widely used system, although it does not compensate for background signals which vary sharply with wavelength.

Both the Zeeman effect and Smith–Hiefte background correction systems use only a single light source and can compensate for a sharply varying background, although they lead to a loss of sensitivity and a fairly limited working range.

Objectives

You should now be able to:

- describe the causes of background interference in AAS.
- explain the basic principles of the techniques available for background correction, and describe their relative merits.

7. Atomic Emission Spectroscopy

Overview

We have seen that most of the elements in the periodic table can be analysed by atomic absorption spectrometry, with varying degrees of ease. So why bother with other techniques such as atomic emission? Emission techniques such as flame atomic emission spectroscopy (flame-AES) and plasma emission spectroscopy (PES) are widely used, and practising analysts would not use them if there were not distinct advantages over absorption techniques.

There are two main advantages of these emission techniques, one specific and the second more general. The specific advantage is that some elements can be analysed with greater sensitivity, or with greater freedom from interferences by emission spectroscopy as compared with absorption spectroscopy. The second, general, advantage follows because the sample is itself the light source in an emission experiment – this means that several elements can be analysed simultaneously, so multi-element analysis offers a tremendous saving in analysis time. Since time is money in an analytical laboratory, purchasing an emission spectrometer capable of the rapid determination of several elements simultaneously can turn out to be cost-effective, even if the multi-element atomic emission spectrometer is much more expensive to buy than an atomic absorption spectrometer.

In this section we shall look first at the use of flame-AES, since the instrumentation has much in common with that in atomic absorption spectroscopy. We shall then look at the other, non-flame, methods of generating the atomic emission. The most popular of these is the inductively-coupled plasma atomic emission spectrometer (ICP-AES) technique which will be studied in more detail than other techniques.

7.1. REVISION OF THE PRINCIPLES OF ATOMIC EMISSION SPECTROSCOPY

Since we have concentrated on *absorption* methods for most of the Unit so far, it may be helpful to refresh our memories with the basic differences between absorption and *emission* methods. The two types of transition are shown in Fig. 7.1a.

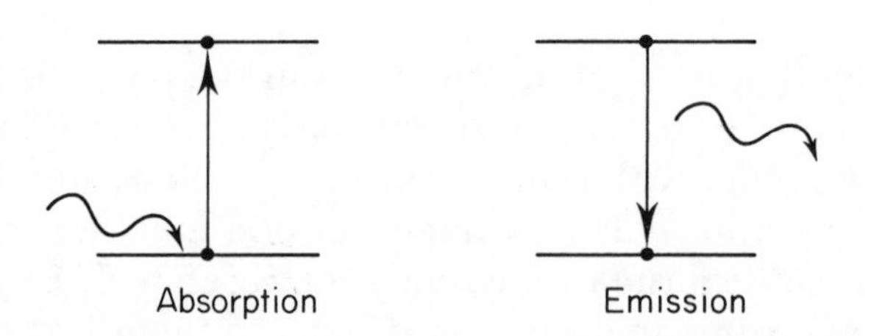

Fig. 7.1a. *Absorption and emission*

Try this revision question to check whether you have understood the principles covered in Part 2 of the Unit

Π In atomic emission spectroscopy which of the following analysis parameters will give the greatest sensitivity?

(*i*) Short wavelength.

(*ii*) Long wavelength.

(*iii*) Low temperature.

(*iv*) High temperature.

(*ii*) and (*iv*) are the correct answers. As we saw in Part 2, the intensity of emission, and hence the sensitivity, depends on the population of the *emitting* state, N_1. The Boltzmann equation (Eq. 2.1)

$$N_1 / N_o = \exp(- \Delta E / RT)$$

$$= \exp(- hcN_A / \lambda)$$

shows that N_1 increases with increasing temperature, and decreases as the energy gap ΔE gets larger. The energy gap is inversely proportional to wavelength, so that N_1 is greater for long wavelength transitions.

Π Which of the following flames do you think will provide the most sensitive method for analysis by atomic emission spectroscopy?

(*i*) Air-propane.

(*ii*) Air-acetylene.

(*iii*) Nitrous oxide-acetylene.

The correct answer is (*iii*).

The nitrous oxide-acetylene flame is the hottest of the three flames listed. Therefore it will give the highest excited state population and the highest emission intensities.

The temperature is very important in emission spectroscopy, since the emission intensity depends so strongly on temperature. This, as we shall see, is one of the reasons for the development of plasma methods which generate much higher temperatures than flame methods.

Π Which type of spectrum for a particular element will contain more lines at a particular temperature – absorption or emission?

Emission spectra are far more complex than absorption spectra. This is because absorption is only from the ground state, but emission can occur not just to the ground state (the exact reverse of the absorption process), but also to a wide variety of other excited states. This was illustrated in Fig. 2.1a and 2.1b.

It is the very complexity of emission spectra which turns out to be one of the major problems with AES. In addition to the emission lines from the excited atoms, emission lines from excited atomic *ions* are also observed in the higher temperature emission sources we shall describe in this Part of the Unit.

7.2. FLAME EMISSION SPECTROSCOPY

There are two commonly used flame emission techniques. One is a specialised instrument, mainly for the analysis of the alkali metals. The instrument is popularly known as a 'flame photometer', and although the term was discouraged by the International Union of Pure and Applied Chemistry (IUPAC), this name is still widely used. The other common flame emission technique is simply an atomic absorption spectrometer operating in the emission mode.

The methods of sample preparation for analysis are essentially the same as for flame-AAS analyses, described in Part 4 of the Unit. This is hardly surprising since both flame-AAS and flame-AES use the same atom cells, and the problems of sample preparation are related to the formation of the atom cell.

Since much of the instrumentation is the same for absorption and emission, we shall concentrate only on those aspects of flame-AES that differ from flame-AAS.

7.2.1. The Flame Photometer

If you take a glass rod, handle the end of it and then put the end of the rod into a roaring Bunsen flame, you will notice the characteristic orange-yellow emission from atomic sodium in the flame. This is caused by the contamination of the rod by sodium chloride present on your hand, due to perspiration. (Even without handling the rod you may see a weaker emission caused by the sodium present in the glass.)

The important point about this observation is that even at the relatively low temperatures of a Bunsen flame (about 2000 K), the alkali metals are significantly atomised, and further excited to a higher energy level. The flame photometer exploits this fact, and consists of a simple arrangement such as that shown in Fig. 7.2a. A Bunsen burner, burning simply mains gas and compressed air, is used to generate the atom cell. Since the emission lines of the alkali metal atoms are well separated:

Li	670 nm
Na	589 nm
K	766 nm
Cs	852 nm
Rb	780 nm
Sr	461 nm

and few other elements emit light strongly at the temperatures of a Bunsen flame, there is no need to use an expensive monochromator. Filters are used to isolate the lines for a particular element. The detector can be a simple photovoltaic cell.

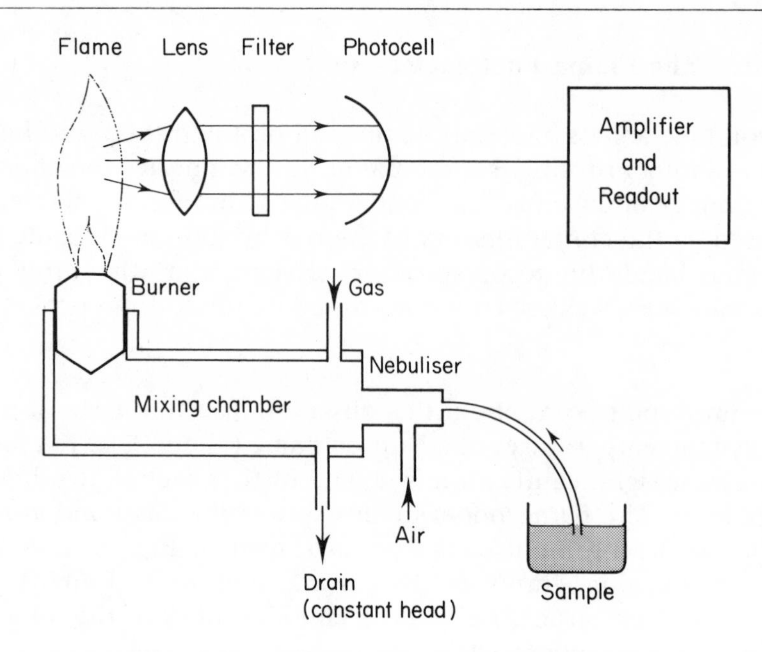

Fig. 7.2a. *The flame photometer*

The technique is suitable for analysing sodium and potassium at 1 to 10 mg dm^{-3} levels. The alkaline earth metals may also be analysed, but with a reduced sensitivity, as shown by Fig. 7.2b. Barium, for example, is usually analysed with a top standard of 1000 mg dm^{-3}, so that only fairly high levels can be determined.

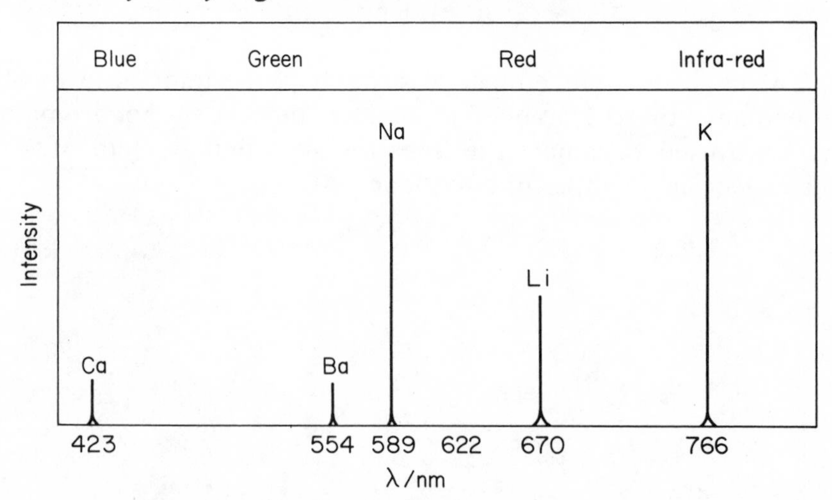

Fig. 7.2b. *Emission intensities at equal concentrations*

For calcium the main atomic emission line is at 423 nm, but the more sensitive emission from $CaOH^+$ at 622 nm is used. Beryllium can be determined using the molecular emission from BeO at 471 nm.

The flame photometer is mainly used for the analysis of clinical samples for sodium, potassium and calcium. The instrument has several advantages:

(*i*) Cost. The flame photometer, because of the simplicity of its construction, is far cheaper than any other AAS or AES instrument.

(*ii*) Freedom from spectral interferences. The low temperature of the flame used means that only a few emission lines are obtained from the flame.

(*iii*) Freedom from ionisation interferences. Since the degree of ionisation of the alkali metals increases with temperature, there will be much less ionisation at the lower temperatures of the Bunsen flame.

The technique does have the disadvantage of only being useful for a few elements. Where the rather poor sensitivity is adequate, flame photometry can be a very cost-effective method.

7.2.2. Modified Atomic Absorption Spectrometers

Many AAS instruments can be simply switched between absorption and emission modes (see Part 3). The best flame is the nitrous oxide-acetylene one, as this gives the highest temperature, and therefore the most intense emission.

Apart from the matrix and ionisation interferences which are common to both flame absorption and flame emission spectroscopy, there are three specific problems which arise in emission – self-absorption, spectral interferences and background emission. (We shall see that these problems are not common to all emission techniques – the high temperature plasma methods are substantially free from many of the above problems.)

(*i*) *Self-absorption*

Fig. 7.2c shows a typical atomic emission calibration curve. A linear relationship between emission intensity and concentration is found at lower concentrations. This is because the intensity of emission is directly proportional to N_1, the number of excited atoms, which is in turn proportional to the sample concentration. At higher concentrations there is a characteristic flattening of the curve. This is due to *self-absorption.* If the emitted photons are from resonance transitions, (that is, back to the ground state), there is a high probability that the photon will be re-absorbed by ground state atoms in the light path. This is one of the main problems in flame-AES, limiting the range of concentrations which can be studied. One of the reasons why the flame technique has not been widely used for simultaneous multi-element analysis is that not all the elements in a given sample are likely to give results on the linear portion of the calibration curve.

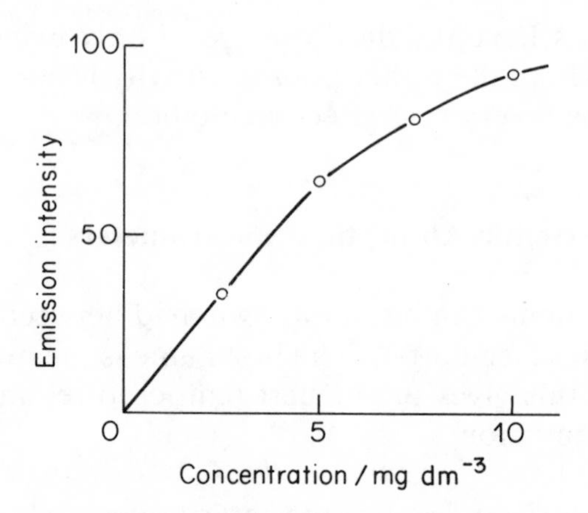

Fig. 7.2c. *Flame-AES calibration curve*

As the next exercises show there are ways of reducing the effects of self-absorption.

Π Will non-resonance lines be absorbed as strongly as resonance lines?

The correct answer is no.

Atoms emitting resonance lines fall back to the ground state, whereas atoms emitting non-resonance lines fall back to an intermediate excited state as shown in Fig. 7.2d. Since the concentration of atoms in the intermediate excited state will be many orders of magnitude lower than the concentration of ground state atoms, the chances of re-absorption of non-resonance lines are negligible.

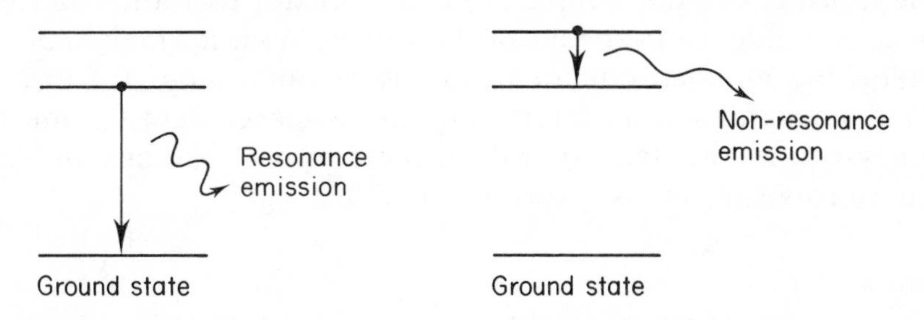

Fig. 7.2d. *Resonance and non-resonance emission*

So we can see that one way of reducing the curvature on the calibration curve, and allowing emission measurements to be made at higher sample concentrations, is to use a non-resonance emission line.

Π Will the amount of self-absorption increase or decrease as the path length through the flame is increased?

Self-absorption increases with path length.

Use of a shorter path length will reduce self-absorption. The slot burners used for AAS are of fixed geometry for a given type of flame. The geometry is determined by the flow velocity requirements of the flame. The nitrous oxide-acetylene flame uses a 50 mm length slot, so to shorten the path length a different burner design can be used. For example, a circular burner such as the Meker burner will give rise to less self-absorption. The Meker burner is essentially a Bunsen burner modified to prevent the problem of flashback - this is done by having a number of small holes rather than one large hole for the gases to pass through.

(*ii*) *Spectral interferences*

Emission spectra are far more complex than absorption spectra. This is because, as we saw near the beginning of the Unit emission is observed for a large number of energy level changes ie to intermediate excited states as well as the ground state. The chances of a near-coincidence of the emission line resulting from another element present in the sample are much greater than the chances of a near-coincidence in the absorption lines. With normal AAS instruments the monochromator band-pass is quite large, 0.1 to 2.0 nm. This means that if an interfering line is closer than 0.1 nm to the analysis line, the detector will respond to both the analyte and the interfering species, as shown in Fig. 7.2e:

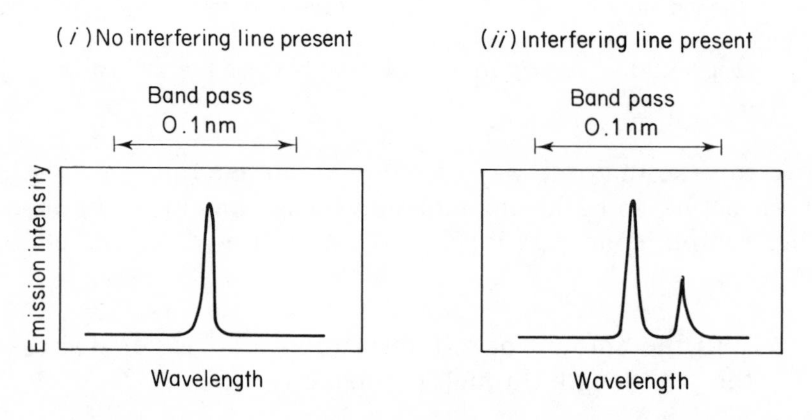

Fig. 7.2e. *Spectral interference of emission lines*

The problem of spectral interference is particularly bad when transition metals such as iron are present as these give large numbers of emission lines (up to 4000 in the uv and visible regions).

One way round the problem is to find another emission line for the analyte where there is no interference.

A second way round the problem is to use a monochromator with a higher resolution. This is not always practical with conventional monochromators which would need to be very long to give enough dispersion, and transmit very little light due to the very narrow slits needed. An alternative is to use an *echelle* monochromator, which has a very high resolution. Eq. 3.8 gave the resolving power, R, as

$$R = nN$$

where n is the order of diffraction and N is the total number of lines in the grating. Values of n of about 100 are used.

Π Which will give the higher resolution – the first order diffraction or the one-hundredth order diffraction?

The one-hundredth order diffraction will give a very much higher resolution than the first order diffraction – by a factor of 100 since R is directly proportional to n.

The echelle grating monochromator uses this principle by working at very high orders of diffraction. The problem with this is that the various orders are superimposed giving a very complex output. Fortunately placing a prism in front of the grating allows the various orders to be displaced.

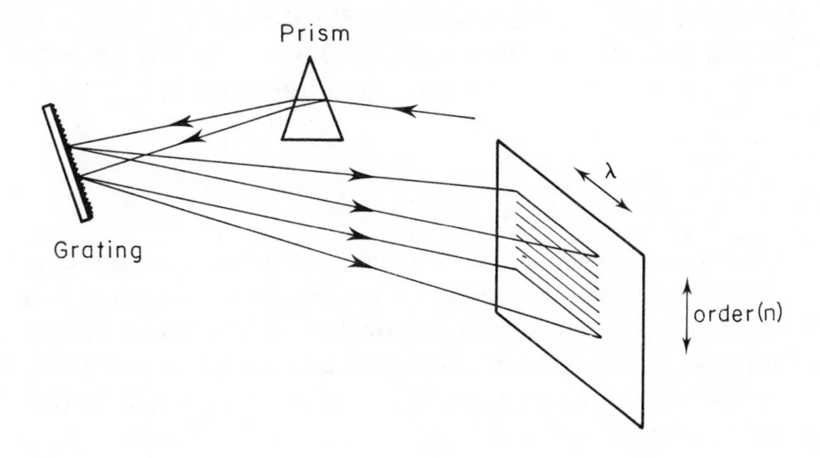

Fig. 7.2f. *The echelle grating monochromator*

The resulting output is seen as a two-dimensional pattern as shown in Fig. 7.2f. The orders change in the vertical direction, and the wavelength changes within a particular order along the horizontal direction. The origin of the word 'echelle', which comes from the French and means 'ladder', should be apparent from Fig. 7.2f.

The resolution of an echelle monochromator is about ten times that of a good conventional monochromator of similar size. Spectral interferences such as that of the 341.26 and 341.23 nm lines of cobalt on the nickel 341.48 nm line are readily avoided, and with care the samarium (492.41 nm) interference on neodymium (492.45 nm) can be eliminated.

The advantage due to the greater resolution of echelle over conventional monochromators, is limited by the line width of the atomic emission lines – there is no point in trying to resolve lines closer than the spectral line widths of 0.003 to 0.01 nm.

(iii) Background correction

As we saw in Part 4, the emission spectrum of a flame contains several molecular emission peaks due to species such as CH, CN, C_2 and OH, superimposed on a broad structureless background. As long as the introduction of the sample does not affect the background emission, then background correction is simple - the reading obtained with the blank solution is subtracted from the reading obtained with the analyte solution. This situation is described in Fig. 7.2g (*i*). In this case the true signal intensity, I_s, is given by

$$I_s = I_1 - I_0$$

Often an extra background signal is obtained from the sample, especially if an organic solvent is used, due to the emission from excited molecular species during combustion. If the background is reasonably flat, then if a wavelength scan attachment is available, a 2-point correction can be performed, as described in Fig. 7.2g (*ii*). The signal intensity is then

$$I_s = I_1 - I_2$$

where I_1 and I_2 are the emission intensities measured at wavelengths λ_1 and λ_2.

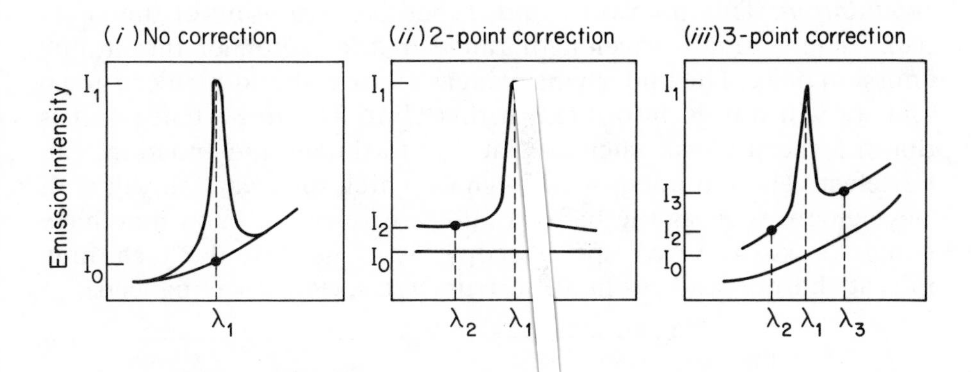

Fig. 7.2g. *Background correction in flame-AES*

If the background is not flat, as shown in Fig. 7.2 (*iii*), the 2-point correction is inadequate and the 3-point correction must be used. The emission intensities I_2 and I_3 are measured on either side of the atomic emission line.

Π Can you work out from Fig. 7.2g (*iii*) which of the following is the corrected signal intensity?

(*i*) I_1-I_0

(*ii*) I_1-(I_2+I_3)

(*iii*) I_1-(I_2+I_3)/2

(*iv*) I_1-(I_2-I_3)/2

The correct response is statement (*iii*).

To obtain the true background emission signal we have to take the average of the two background signals which are measured on either side of the analysis wavelength. The background at the atomic emission line is then $(I_2+I_3)/2$ and this must be subtracted from I_1 to obtain the corrected signal.

An alternative method for background correction is *wavelength modulation*. This involves rapid, repetitive, scanning of the spectrum over a small wavelength range in the region of the atomic emission peak. The underlying principles are rather difficult to grasp and we will not go into them in this Unit. The important point is that the sharp peaks, such as that due to the atomic emission, can be selectively amplified, while signals which only vary slowly with wavelength, such as the background, are rejected. (You may have come across derivative spectroscopy, which is a similar technique in that sharper features in a spectrum are selectively enhanced).

SAQ 7.2a

Which of the following statements is correct?

(*i*) Flame emission spectroscopy is not subject to the same matrix interferences as flame-AAS.

(*ii*) Spectral interferences can be reduced by using a smaller spectral bandpass for the monochromator.

(*iii*) 2-point background correction is satisfactory for strongly sloping background emission.

7.3. NON-FLAME EMISSION SPECTROSCOPY

7.3.1. Limitations of Flame Emission Spectroscopy

There are three types of interference in atomic spectroscopy, all of which can occur in flame-AES. These are:

(*i*) Chemical interferences, also known as matrix interferences. As we saw in Part 4, the degree of atomisation of the sample in the flame is affected by the chemical nature of the sample.

(*ii*) Ionisation interferences. The degree of ionisation of some easily ionisable elements, and hence the degree of atomisation, is affected by the presence of other easily ionisable elements.

(*iii*) Spectral interferences. The presence of spectral lines of other elements present close to the spectral line of the analyte can give misleading results.

The chemical and ionisation interferences will be similar for both absorption and emission flame spectroscopy, while spectral interferences are a far greater problem with emission.

In addition, as we saw in the previous Section, self-absorption and background emission can also be problems.

It is probably true to say that flame-AES is an under-used technique. This is because there are potentially more interference problems in flame-AES than in flame-AAS, and although there are some elements for which greater sensitivity can be obtained by flame emission, these elements can usually be adequately analysed by absorption methods.

7.3.2. High temperature non-flame emission sources

Another reason for the limited use of flame-AES is that the more recent development of higher temperature non-flame emission sources, has tended to divert attention from the use of flame emission sources. These sources are usually *plasmas* and can operate at temperatures of up to 10 000 K or so. The most immediately obvious advantage of using higher temperatures is the greater concentration of emitting atoms, particularly for those atoms with emission lines in the ultra-violet. The effects of higher temperatures on the various interferences are also important. Perhaps you can work out what the likely effects are

Π Will the three types of interference listed below be reduced or increased by raising the temperature of atomisation?

(*i*) Ionisation.

(*ii*) Chemical.

(*iii*) Spectral.

(*i*) At least at first sight we would expect ionisation interferences to be worse at higher temperatures because more atoms will be ionised. In practice, however, ionisation interferences are not always very severe in plasmas, probably due to the very high electron densities present in the plasmas. (The extra electrons released by an easily ionisable element will have a negligible effect on the ionisation equilibria of other elements, since the extra electrons released form only a small fraction of the total electron concentration in the plasma.)

(*ii*) Chemical interferences will be reduced as the temperature is raised. This is because chemical interferences are due to varying degrees of ease of atomisation. For example the well-known suppression of calcium signals by phosphate is due to the reduced efficiency of atomisation when relatively stable molecules can be formed. At the higher temperatures of plasmas, only very small concentrations of molecular species are present, so chemical interference are not a problem.

(*iii*) Spectral interferences will be increased at higher temperatures. Since more atoms will be in very highly excited states, there will be an increase in the number and intensity of non-resonance emission transitions (ie transitions to lower-lying excited states rather than the ground state).

At first you might think there is something of a trade-off in working at higher temperatures – you get rid off chemical interferences, but replace them with spectral interferences. This is true, but there is an overall gain. Spectral interference can usually be identified and avoided by suitable instrumental adjustments. With modern computer-controlled instruments this process is normally straightforward. Chemical interferences on the other hand, can be reduced to some extent in flame spectroscopy by optimising the instrumental technique (eg using high temperature fuel-rich flames for barium to minimise BaO formation), but it is often necessary to carry out time-consuming analyses involving sample pre-treatment, such as the addition of a releasing agent, or using the method of standard additions, etc.

So although we are replacing one type of problem by another, the high temperature problem (spectral interference) can usually be easily controlled, whereas the low temperature problem (chemical interference) tends to be more time-consuming to solve, and more dependent on the skill of the analyst.

As we shall see there has been a great deal of interest in high temperature emission sources, for four main reasons.

(*i*) The freedom from chemical interference.

(*ii*) The greater sensitivity, particularly for ultra-violet emission lines.

(*iii*) The long linear working range (about 4 orders of magnitude of concentration).

(*iv*) The potential for simultaneous multi-element analysis.

The last point is of course true for all emission techniques, including flame-AES, but there has not been the same degree of interest in multi-element flame-AES as in multi-element non-flame emission spectroscopy. This is probably because the developments in computing, needed to handle large amounts of data rapidly, have coincided with the developments in the non-flame methods.

7.3.3. Types of non-flame emission source

There are several types of high temperature sources, but the most popular sources are plasmas. A *plasma* is essentially a highly ionised gas - definitions vary slightly, but generally a gas is considered to be a plasma if about 1% or more of the atoms in the gas are ionised. The plasma is obtained by subjecting the gas to an electrical discharge. The electrical discharge can be obtained in various ways. The most widely studied techniques are:

(*i*) the direct current arc plasma, in which a high constant current is passed through argon gas between two (or three) electrodes;

(*ii*) the inductively coupled plasma, in which a high frequency alternating current (usually 15–50 MHz) is used to form a plasma in argon; and

(*iii*) the microwave-induced plasma, in which an ultra-high frequency alternating current (2450 MHz) forms a plasma in helium.

These are not the only methods of generating high temperature plasmas. *Sparks*, both DC and AC have been used as emission sources for some time. The potential of *lasers* for generating suitable emission sources is also being investigated.

7.4. DC PLASMAS (DCP)

The DC plasma is basically an arc discharge between two electrodes. The sample is nebulised at a flow rate of about 1 $cm^3\ min^{-1}$ in an argon carrier gas and injected into the discharge. The two-electrode arc has more recently been superseded by a three-electrode design as shown in Fig. 7.4a. The third electrode helps to stabilise the discharge which has a characteristic Y-shape.

For most elements the detection limits are no better than for flame-AAS. Very often they are worse, but some of the more difficult elements such as boron and phosphorous, can be determined at much lower levels with the DCP by virtue of the higher temperatures of the plasma. The plasma temperatures are high (up to 10 000 K), but measurements are made just below the plasma, at about 5500 K, due to the intense background emission from the plasma.

Although the DCP is more economical than ICP spectrometers (Section 7.6), the DCP is less widely used. This is for several reasons:

(*i*) detection limits are not so good – usually about a factor of 5 or 10 worse than the ICP;

(*ii*) ionisation interferences are more of a problem with the DCP than with other plasmas. Emission intensities are increased when easily ionisable elements are present in the sample.

(*ii*) as with other plasmas, sample introduction is a limiting factor in the precision which can be obtained.

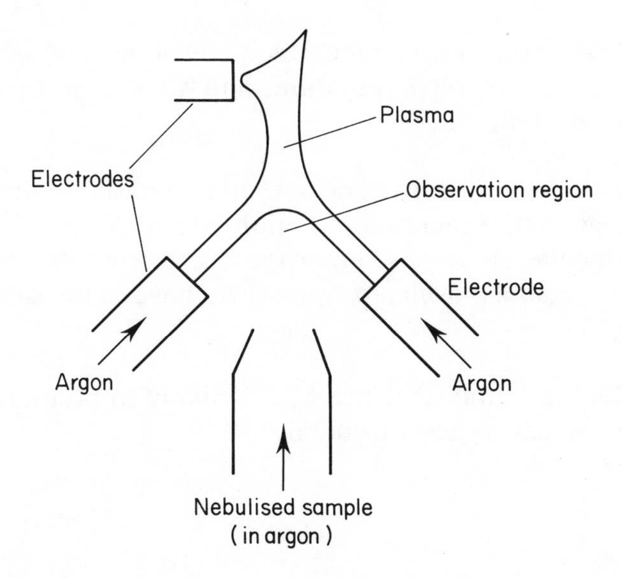

Fig. 7.4a. *Three electrode direct current plasma*

7.5. THE MICROWAVE-INDUCED PLASMA (MIP)

Microwaves are a very high frequency alternating electrical current. A 200 W power supply at a frequency of 2.45×10^9 Hz is used to generate the plasma, because medical diathermy units operating at this frequency are readily available. The temperature of this plasma is especially difficult to define, because it is not in thermal equilibrium. The gas temperature is only about 1000 K, but much greater concentrations of excited atoms are present than would be predicted if the plasma was at equilibrium. The excited atoms correspond to an excitation temperature of perhaps 7000–9000 K and are formed in collisions with metastable helium atoms. (A metastable atom is in a relatively long-lived excited state). The following reactions are a simplified version of what is thought to occur in the plasma.

$$He^* + X \rightarrow He + X^+ + e^-$$

$$e^- + He + X^+ \rightarrow He + X^*$$

$$X^* \rightarrow X + h\nu$$

An asterisk indicates an excited state. The excited atom is formed from electron-ion recombination, and so will be in a very highly excited electronic state.

The reason for using helium is that a metastable argon atom has less energy than a metastable helium atom. Argon plasmas tend to give molecular emissions rather than atomic emissions. In order to maintain a stable plasma, low pressures have to be used (5 to 10 mm Hg).

Π Do you think that the MIP is likely to be useful for the determination of non-metals?

Yes, the MIP can be used to determine non-metallic elements.

The high excitation temperature will favour the formation of emitting states of non-metal atoms, as these atoms need much more energy to populate excited states than metal atoms. We covered this point in Part 1 of the Unit.

Another way of thinking of this is that since the atoms are ionised in the discharge, then the recombination process *must* form highly excited atomic states. On the energy level diagram for an atom the highest energy level is always the ionisation limit (Fig. 1.4a). Many emission lines from highly excited states to less energetic excited states will be in the uv-visible and will be readily detectable.

The main advantages of the MIP are that it is cheap (relatively) and the high excitation temperature allows non-metals to be detected, although the detection limits are not especially low.

The main disadvantages are the poor detection limits, the plasma is easily quenched by water (so that the introduction of aqueous samples is a problem), and chemical interferences can be a problem. These problems are largely due to the low power of the plasma.

The MIP then is most suited to applications where the sample is injected as a gas. A very useful application is as a selective detector for gas chromatography. In a typical configuration up to twelve elements can be monitored simultaneously. So if a gas chromatograph is used to separate a complex mixture of compounds, the elemental analysis of each compound can be carried out as it is eluted from the chromatograph. A schematic representation of such a detector is shown in Fig. 7.5a.

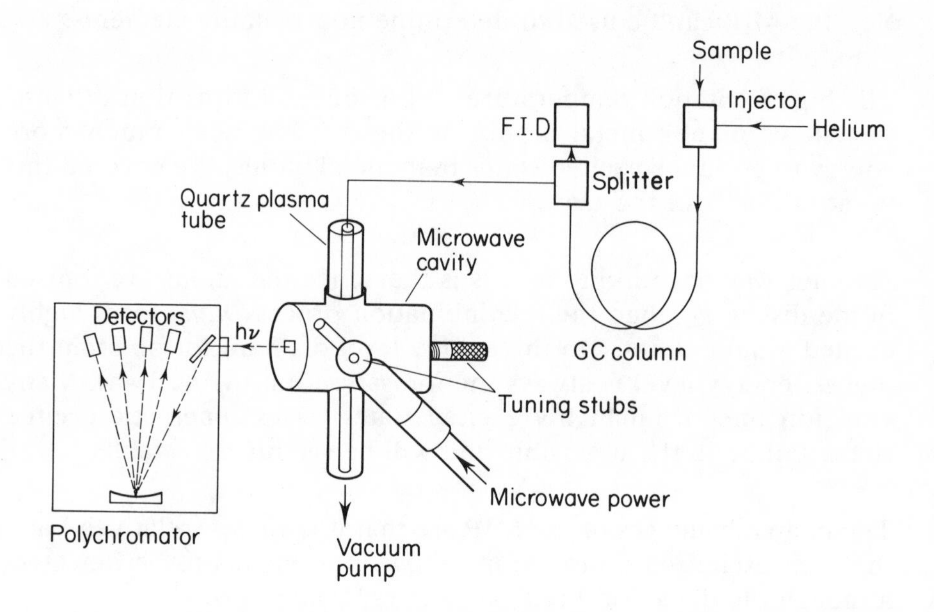

Fig. 7.5a. *The MIP as a simultaneous multi-element gc detector (schematic)*

The mixture to be analysed is injected as a gas or a solution into the chromatograph, to be flushed along a separating column with helium gas. After separation some of the column effluent passes to a conventional gc detector, usually a non-selective FID (flame ionisation detector). The rest of the effluent is pumped at reduced pressure through a quartz plasma tube which passes through the microwave cavity. The light emission is directed from a hole in the centre of the microwave cavity onto the entrance slit of a *polychromator*.

A polychromator is a grating monochromator in which the grating is fixed and photomultipliers are positioned at the wavelengths of the most intense spectral lines of several elements. Polychromators are used for all the *simultaneous* multi-element emission methods which are described in this Part of the Unit.

The arrangement shown in Fig. 7.5a uses a concave grating which both disperses and focuses the light. You may notice that the detectors are arranged in an arc rather than in a straight line – they are located on the *Rowland circle* of the grating, along which the different wavelengths are focussed. This type of grating was originally developed by Rowland in 1882. The principle of the Rowland grating is described in Fig. 7.5b. The important point is that the diameter of the circle on which both the entrance slit and the detectors are placed is equal to the radius of curvature of the concave grating.

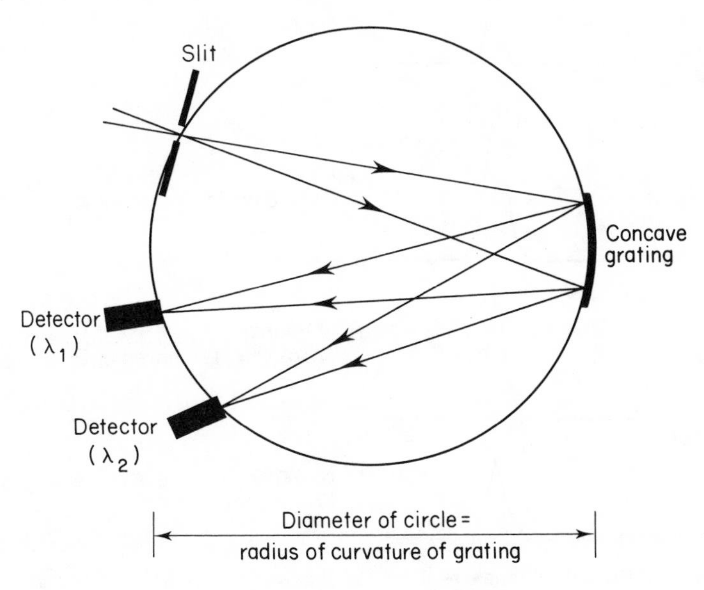

Fig. 7.5b. *The Rowland circle*

Clearly if we have a large stack of pen recorders, or better, computer-acquisition of the data, we can simultaneously analyse as many elements as we like provided that we can afford the necessary photomultipliers and find space for them all in the polychromator.

A simple application is shown in Fig. 7.5c. This shows the results obtained when a simple two-component mixture is separated on the chromatograph. The gc detector trace (*i*) shows the two components as peaks A and B, but gives us little information, apart from the retention time, about the identity of compounds A and B. The MIP as a detector allows us to determine the empirical formula of the compounds, provided that we have an internal standard. This is because the area of a given element peak is proportional to the number of atoms of the element present. Traces (*ii*)–(*iv*) show the responses for C, H and N channels from the polychromator.

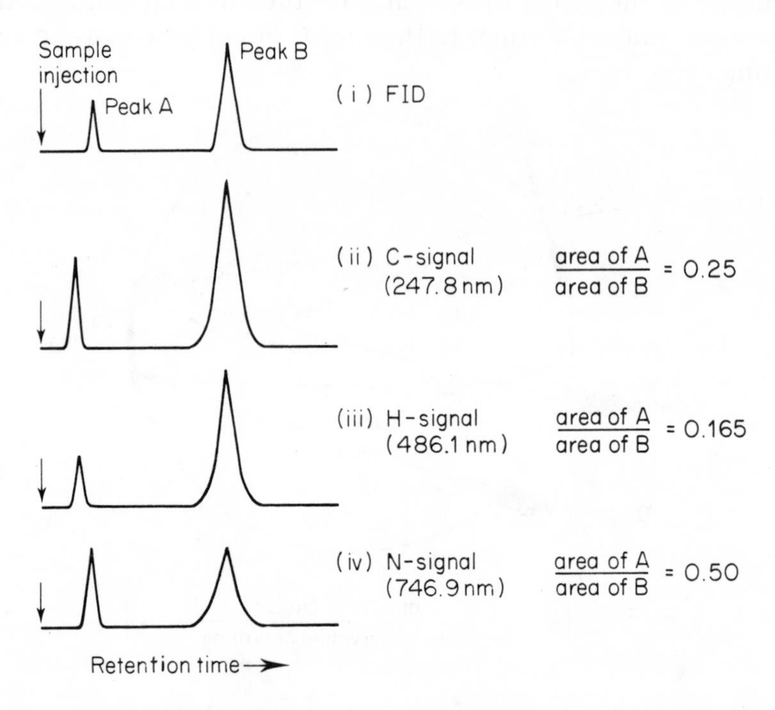

Fig. 7.5c. *Typical output from a MIP used as a gc detector (with FID trace for comparison)*

Π In the example shown peak B is due to acetonitrile, CH_3CN. Can you identify peak A from the ratios given for the peak areas on the nitrogen, carbon and hydrogen traces?

Compound A is hydrogen cyanide, HCN. The empirical formula for acetonitrile is $C_2H_3N_1$. The empirical formula of compound A is then:

$$C_2 \times 0.25 \qquad H_3 \times 0.165 \qquad N_1 \times 0.50$$

This simplifies to $C_{0.50}\,H_{0.495}\,N_{0.50}$ or CHN, better known as HCN.

Note that the numbers in the calculation do not round up to exact whole numbers, and you might think at first that this might lead to ambiguities in converting empirical into molecular formulae. This is not a great problem in practice because we can usually get a good idea of the number of carbon atoms in a compound from the gc retention time.

The MIP is a particularly useful gc detector in environmental analysis where highly complex mixtures are often analysed, as rapid identification of the compounds of interest is possible. For example pesticide residues are often chlorinated aromatics, and these can be readily monitored using the chlorine atom emission from the plasma.

A limitation to the usage of the MIP as a gc detector is popularity of the use of mass spectrometers to carry out direct analysis of the gc effluent. The mass spectrometer gives far more information about a compound than the MIP, and can usually allow the chemical structure to be determined as well as the empirical formula. On the other hand the MIP is cheaper and simpler than most mass spectrometers, and may still be the method of choice for some applications.

7.6. THE INDUCTIVELY COUPLED PLASMA

The inductively coupled plasma (ICP) is also known as the radiofrequency (RF) plasma. This is operated at high power levels, 0.5 to 3.0 kW and a frequency of 15–50 MHz. Temperatures over 10 000 K are reached in the plasma. The basic principle is illustrated in Fig. 7.6a. The radiofrequency electrical current is passed through a metal induction coil. The current has an associated magnetic field, with lines of force passing along the axis of a quartz tube placed inside the coil. Electrons are accelerated by the electromagnetic field to travel in circular orbits inside the quartz tube. Energy is transferred from the electrons to the gas by collisions and so the gas heats up. The temperatures reached produce high concentrations of both excited atoms and ions.

Although we shall concentrate on atomic emission lines it is important to remember that many (atomic) ionic emission lines are also present, and these can also be used for analysis.

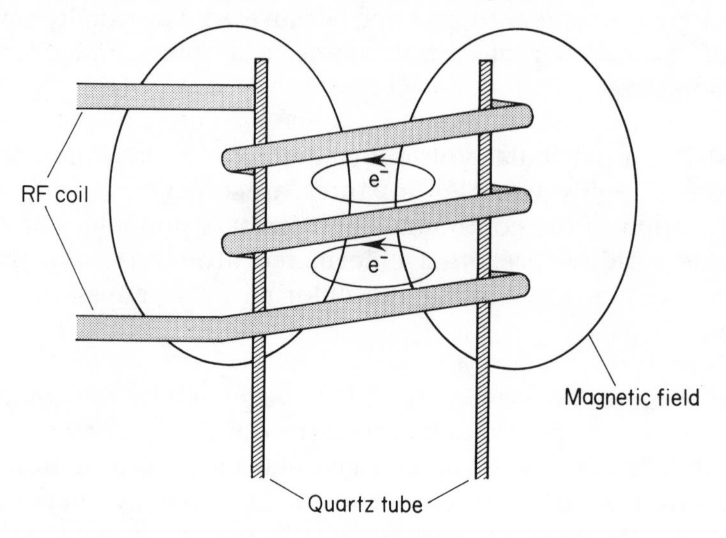

Fig. 7.6a. *The inductively coupled plasma*

Π Which requires the higher temperature – the formation of an excited atom or the formation of an excited ion?

The excited ion is formed at higher temperatures than the excited atom.

More energy is required to ionise an atom than to form an atomic excited state, since the highest energy level of an atom is the ionisation limit. Even more energy will be required to excite the ion, so that the formation of excited ions needs higher temperatures than the formation of excited atoms.

In the typical arrangement of a practical ICP unit, known as an ICP torch, and described in Fig. 7.6b, three gas flows along three concentric tubes are required. In an all-argon unit the plasma is maintained by flowing argon through all three tubes. The sample is drawn up as an aerosol in argon into the centre of the plasma. The second gas flow maintains the plasma. The third, coolant, argon (or nitrogen) flow is required around the outside of the plasma to prevent the quartz tube melting. Large amounts of gas (about 10 $dm^3 min^{-1}$) are used in most current designs.

The plasma has a characteristic doughnut shape with the sample being introduced into a relatively cool (6000–8000 K) hole in the centre of the doughnut (see Frontispiece). The light emission from the white hot fireball of the plasma is very intense, so analytical measurements are made in the cooler tail plume of the plasma. The spectral background in the plume, particularly in the useful 190–300 nm region, is relatively simple, consisting mainly of argon emission lines.

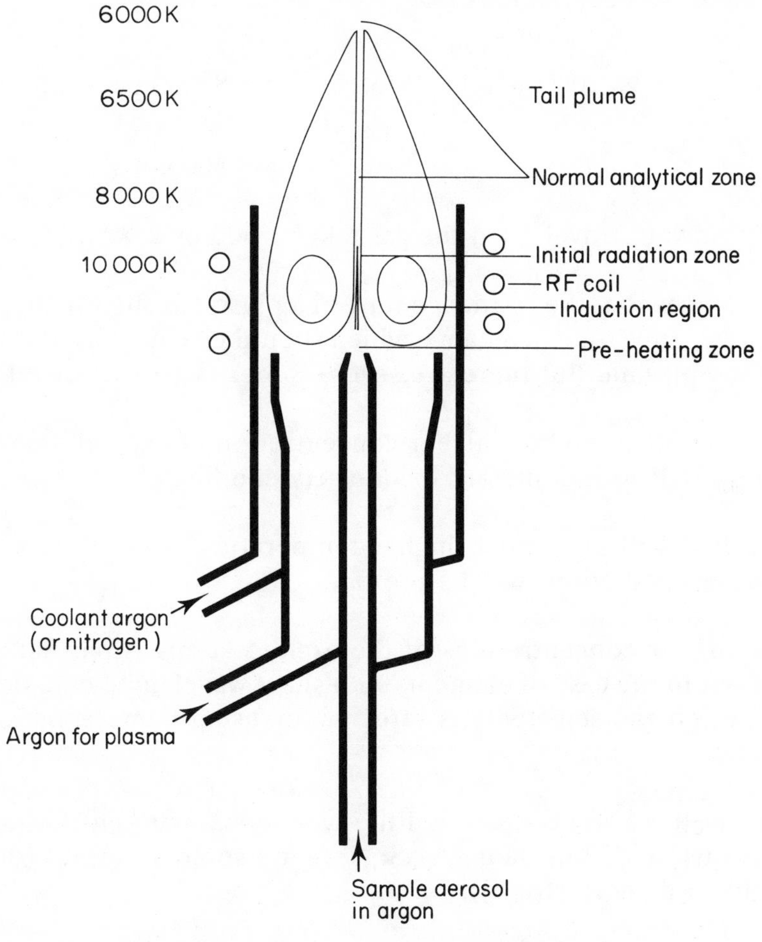

Fig. 7.6b. *An ICP torch*

We can make a rough estimate the residence time from the flow rate.

Π Assume that the total gas flow rate in an ICP torch, allowing for expansion effects, is 180 $dm^3\ min^{-1}$, and that the cross-sectional area of the plasma is 6 cm^2. Can you estimate the time taken by the gas to travel 1 cm through the plasma?

The answer is 2 milliseconds.

The linear flow rate of the gas is simply the volume flow rate divided by the cross-sectional area or

$$1.8 \times 10^5(cm^3 min^{-1})/6(cm^2) = 30000\ cm\ min^{-1}$$

$$= 500\ cm\ s^{-1}$$

Therefore to travel 1 cm the gas takes 1/500 or 2×10^{-3}s.

We can see that the residence time in a plasma is slightly longer than the residence time in a flame, which should give improved sensitivity for the plasma. But there are several other factors involved.

Π Will there be a higher concentration of excited atoms in the ICP or in a nitrous oxide-acetylene flame?

The ICP will give much higher temperatures than flames, so that more excited atoms will be present.

The higher concentrations of the excited atoms is particularly important in the case of elements with short wavelength emission lines, for which the sensitivity is rather poor using flame emission techniques.

For most elements there is little to choose between the detection limits using ICP or flame-AAS. Here are some examples (detection limits in ng cm^{-3} (ppb)):

Element	Flame-AAS	ICP
Al	20	10
B	1000	2
Cd	1	1
Cu	2	2
P	100 000	30
Pb	10	15
Zn	0.6	1

Π From the list of detection limits given, can you name three elements for which ICP gives better sensitivity, and two elements for which flame-AAS gives better sensitivity?

Al, B and P are all detected with greater sensitivity by ICP, while Pb and Zn are detected with greater sensitivity by flame-AAS.

Note that ICP is more suitable for those elements which are particularly difficult to analyse by flame-AAS, since they tend to form relatively stable molecular species at flame temperatures. Atomisation is increased in the plasma because of the longer residence times in the plasma, combined with the higher temperatures.

7.6.1. Sample introduction into the ICP

The sample is usually introduced by solution nebulisation, much as for flame spectroscopy, although we shall see that the nebuliser design is different. Considerable interest is also being shown in both the direct introduction of solids into the ICP and gaseous introduction in the hydride generation-ICP technique.

Sample solution preparation is often rather easier for ICP spectroscopy than for flame spectroscopy since the freedom from chemical interferences means that less care is required in sample pretreatment.

The nebulisers developed for flame spectroscopy have been designed for higher gas flow rates (10–20 dm^3 min^{-1}) rather than for ICP spectroscopic use, where flow rates of only 1 dm^3 min^{-1} of carrier gas or so are needed to take up the sample at about 1 cm^3 min^{-1}. Nebuliser design is one of the most critical parts of the design of the ICP system, since much of the 'noise' in the emission signal is due to nebulisation problems.

Three types of nebuliser are shown in Fig. 7.6c. The first two, the *crossed flow* and the *concentric* nebulisers are the most widely used. Much narrower jets are used than in flame-AAS nebulisers, so that care has to be taken to prevent blockages. In the concentric nebuliser, also called a Meinhard nebuliser, the argon flows out through a narrow circular orifice in the jet, around the sample capillary. The third type, the V-groove or *Babington-type* nebuliser, is useful for handling slurries, since it is far less prone to blockage and can handle slurries containing up to 5 or 10% of solids. The sample solution flows along the bottom of a V-groove and is nebulised by the argon gas which is forced though the solution from a small hole in the groove. Several other ICP nebuliser designs are also being developed.

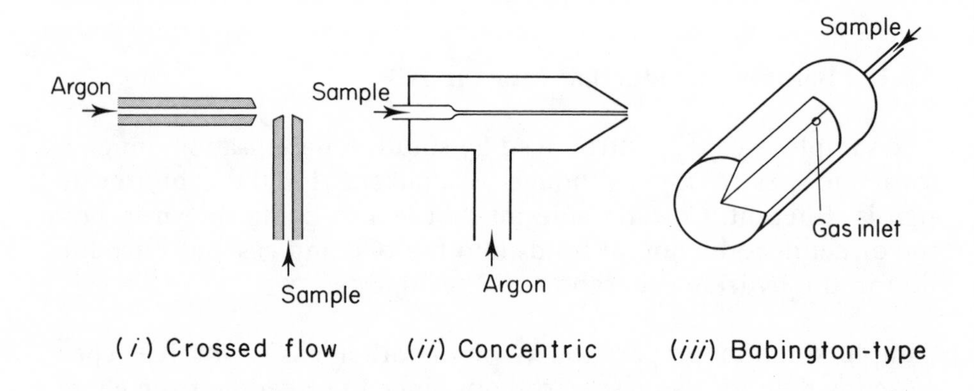

Fig. 7.6c. *Nebulisers used in ICP spectroscopy*

The precision obtained in ICP analysis is usually 1 or 2%, which is better than furnace-AAS but not quite so good as flame-AAS (0.3%). The limitation in precision, due partly to emission noise caused by nebulisation, can be improved to some extent by the use of internal standards. The emission intensity is also very sensitive to plasma temperature, so very stable radiofrequency power supplies are required.

An alternative to the usual *continuous* methods of sample introduction for solid samples is to use *discrete* methods. The sample can be introduced directly on the end of a graphite rod underneath the plasma. Other methods involve vaporising the solid and sweeping the vapour in flowing argon into the plasma. The vaporisation can be carried out using an electrothermal atomiser, a pulsed laser or a high current spark.

7.6.2. Determination of samples

The main group elements can give hundreds of emission lines, while transition metals may give several thousand emission lines in the uv-visible region. It is not surprising that the main problem in ICP spectroscopy is spectral interference, and so an important part of method development for complex samples is to select interference-free lines for the elements of interest. Representative samples that have been analysed by another method are useful, as these can be analysed at several wavelengths, say at 308.22, 394.40 and 396.15 nm for aluminium, and those lines subject to spectral interferences will give high values for the concentrations. Any of the wavelengths giving a correct answer can be used.

An example of spectral interference is shown in Fig. 7.6d. The cobalt peak at 228.616 nm is overlapped by a peak due to chromium. The solid line shows the total emission, and the dotted lines show the contribution from each element.

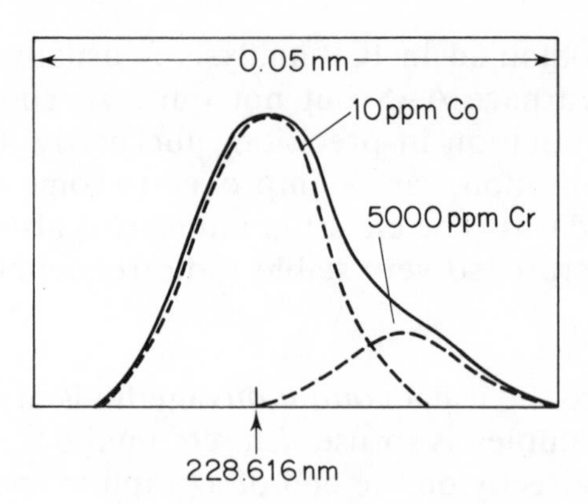

Fig. 7.6d. *An example of spectral interference*

Other methods can be used to deal with spectral interferences. If the interfering element is known, then its concentration can be estimated from another of its interference free emission lines, and this can be used to make a correction to the intensity of the sample emission line. Alternatively since the instrument is normally computer-controlled, there is no reason why curve resolution techniques, which have been developed for other forms of spectroscopy, should not be used. One more obvious method is to use a narrower spectral bandpass for the monochromator.

Π The best resolution obtained with commercially available scanning monochromators is about 0.01 nm. Is this low enough to satisfactorily resolve the two overlapping peaks shown in Fig. 7.6d?

The answer is 'yes' – just. If the monochromator is set to transmit at the peak wavelength of the cobalt emission then there will be a small contribution from the wing of the chromium peak which is also transmitted giving a slightly high reading. This could be avoided by setting the monochromator to transmit a little to the low wavelength side of the cobalt peak.

Note that the two peaks are separated by about 0.02 nm, and they can be just resolved by a monochromator resolution of 0.01 nm.

Peaks closer than 0.01 nm cannot normally be resolved by a scanning monochromator. Bear in mind also that these considerations are affected by the emission line widths – broader peaks are harder to resolve.

Many instruments allow the observation height above the plume to be varied, and a plot of emission signal against observation height, as shown in Fig. 7.6e, will allow the most sensitive position to be found.

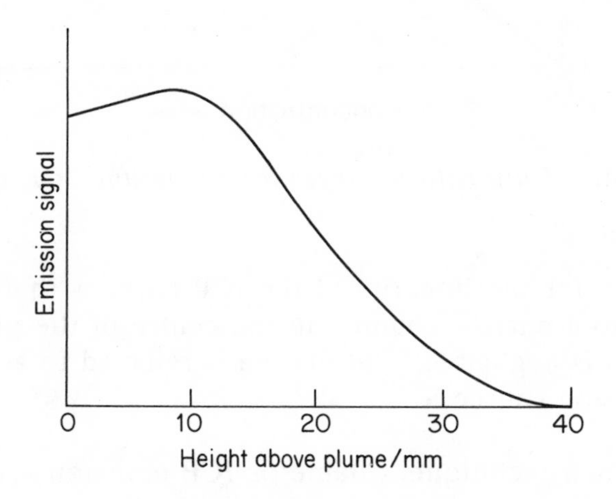

Fig. 7.6e. *Variation of emission signal with plume height*

If background emission is present and a scanning monochromator is used, a 2-point or 3-point correction method, as described earlier in this Section, may be used.

The setting up procedure is then repeated for the next element to be analysed.

Calibration curves are obtained for each element. A typical calibration curve for an ICP analysis is compared with a flame spectroscopy calibration curve (absorption or emission) is shown in Fig. 7.6f.

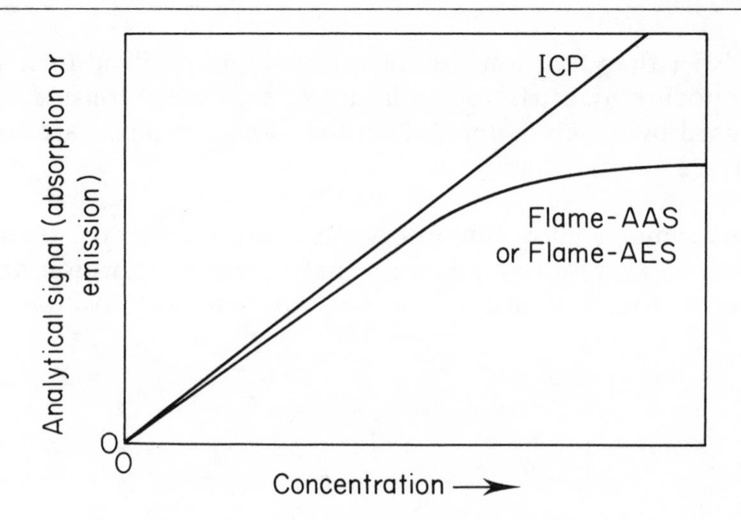

Fig. 7.6f. *Calibration curves for flame and ICP spectroscopy*

The reason for the linearity of the ICP curve is that the sample is confined to a narrow channel in the centre of the plasma, so self-absorption is negligible. The plasma is referred to as an 'optically-thin' emission source.

Π Which technique – flame or ICP emission – is more useful for the simultaneous analysis of both trace and major components in a mixture?

ICP, because of the highly linear calibration curve, can be used to analyse both major and minor components simultaneously. With the flame technique, the major components would be detected in the flat region of the calibration curve. This would greatly reduce the precision of the determination of the major components and could only be overcome by diluting the sample for a separate set of measurements.

7.6.3. Sample throughput

One of the great advantages of an instrument such as a *simultaneous* multi-element ICP spectrometer, is the very high potential sample throughput.

If a series of samples is to be analysed for a high number of elements – 12 is a reasonable number – then the number of determinations that can be made is large. Each sample may take nearly 1.5 minutes to analyse, 1 minute for the sample concentration in the plasma to reach equilibrium and up to 30 seconds to take readings, make background corrections and print out results. Therefore about 40 such samples can be analysed in an hour, so the total throughput is 12 elements × 40 samples or 480 elements per hour.

Although higher sample throughputs can be obtained (3600 has been claimed for 60 elements at 1 sample per minute), particularly if more elements are to be determined in each sample, these numbers are intended to be realistic. It is not often that more than about 12 elements are analysed simultaneously.

An important assumption in this calculation is that all the elements to be determined need to be determined with only moderate precision of about 3 to 5% (higher if the concentration is near the detection limit). If a precision of 1% is needed then a longer analysis time must be used, (Fig. 7.6g).

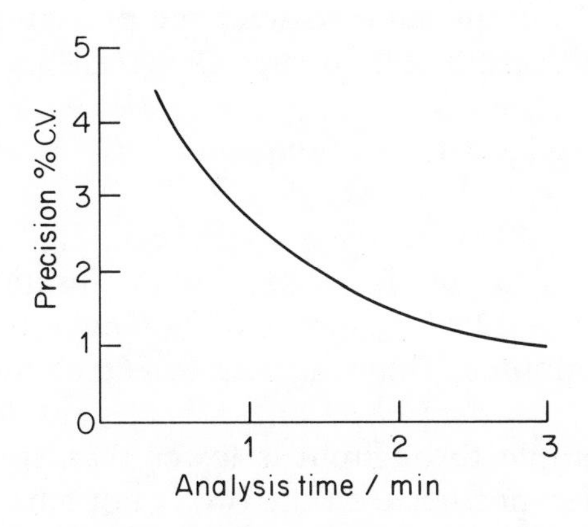

Fig. 7.6g. *Variation of precision (expressed as percent coefficient of variation) with analysis time per sample*

The problem with simultaneous multi-element analysis using a polychromator, is the lack of flexibility. The photomultiplier detectors are fixed in position and it is expensive to change or add photomultipliers to study elements not originally specified, or to avoid spectral interferences by using a different line for an element.

Greater flexibility can be obtained with a *sequential* ICP system, which uses a scanning monochromator instead of a polychromator. For high sample throughput to be obtained the monochromator must be scanned both very rapidly and accurately ($\pm$0.005 nm). These demanding requirements can be met using a computer-controlled monochromator, with a stepper motor drive for the grating. A wavelength calibration is carried out using a low pressure mercury discharge lamp which has well-defined emission lines throughout the uv-visible region. In operation the monochromator, using an internal reference lamp such as a low pressure mercury discharge lamp, starts at a mercury line such as 365.02 nm and then scans the elements in sequence. The analysis time is perhaps 5 seconds for each element – 3 seconds to change wavelength and 2 seconds to integrate the signal.

Π Can you compare the sample throughput for a sequential ICP with that for the simultaneous ICP described above? Assume 1 minute stabilisation time per sample, 12 elements per sample and 5 seconds per element.

The sample throughput for the sequential ICP is 360 elements per hour.

For each sample, 1 minute is needed for stabilisation and 1 minute for measurement – 12 × 5 seconds – giving a throughput of 12 elements per 2 minutes. This scales up to 360 elements per hour.

Although the sample throughput is lower than for the simultaneous ICP, the difference between the two is not particularly marked. If considerably greater than 12 elements were to be analysed then throughput would be much greater for the simultaneous system.

For most applications the sequential spectrometer is attractive because of its greater flexibility, and generally lower cost (only one detector is required). Some instruments capable of both simultaneous and sequential ICP analysis are available, allowing the analyst to have the best of both worlds, at a cost.

It is useful to compare these results with the sample throughput for a single element flame-AAS spectrometer. Sample equilibration is faster and readings can be taken in about 12 seconds, but the lamps have to be changed and the samples re-analysed, for each element.

Π Compare how long it will take to analyse 40 samples for 12 elements by flame-AAS with the time taken by simultaneous ICP (1 hour). Assume that all the samples are analysed for one element at 12 seconds per sample, and that the lamp is then changed in 3 minutes so that all the samples can be analysed for the next element and so on.

The analysis time is 129 minutes.

Each element takes about $12 \times 40 = 480$ seconds or 8 minutes to analyse for 40 samples. 12 sets of analyses will take $12 \times 8 = 96$ minutes, and allowing for 11 lamp changes which will take $3 \times 11 = 33$ minutes, the total analysis time is 129 minutes. This is just over twice as long as the simultaneous ICP.

The calculation is rather simplified in that the time to prepare calibration curves has not been included, and this will take longer for flame-AAS. On the other hand when only a few elements are to be analysed, flame-AAS may be faster than ICP methods, since the equilibration times for measurements in flames are shorter.

An important factor we have not so far considered is the time spent on method development. Very little time is usually needed for development of a flame method, but multi-element analysis ICP methods can take some time to develop.

The estimates of sample throughput in this section should be taken only as being indicative – times will vary for instruments from different manufacturers, and we have ignored various other factors such as sample preparation time, warm-up time for lamps, the possible need to change burners for some flame measurements etc. These factors would tend to lengthen the analysis time for flame-AAS. On the other hand many manufacturers now produce 'multi-element' flame-AAS spectrometers. Typically up to 8 pre-warmed lamps may be operated sequentially by computer control to give higher sample throughput for multi-element flame-AAS analysis.

SAQ 7.6a

Which of the following statements are correct?

(*i*) MIP and DCP methods give lower detection limits than ICP methods.

(*ii*) Flame-AAS is a better method than ICP for the analysis of refractory elements such as aluminium and phosphorus.

(*iii*) A much slower carrier gas flow through the nebuliser is required for an ICP torch than for a air-acetylene flame.

(*iv*) Self-absorption is more of a problem than spectral interferences in ICP spectroscopy.

7.7. DIRECT ANALYSIS OF SOLIDS – ARCS AND SPARKS

A direct current (DC) *arc* is an electrical discharge between two electrodes. The high current of the discharge (10 amps or so at 10–100 volts) can displace atoms from the electrode surface. The atoms become excited by the discharge in the gas and so act as an emission source. A *spark* is a short-lived discharge at higher voltages (several thousand volts).

The sample can be placed on an electrode or evaporated onto it from solution. The most important use, however, is the direct analysis of metal samples by using the sample as the electrode. The *anode* is used as the sample electrode as it gets hotter than the cathode.

The precision of the DC arc is rather poor but can be improved if an intermittent discharge is used. Although sometimes called an *AC spark*, it is more correctly a pulsed (50–500 Hz) unidirectional or monoalternant discharge. This means that the current only flows in one direction so that the sample electrode is always the anode.

Π The average discharge current in an intermittent DC arc is 1 amp. A continuous DC arc operates at a current of 10 amps. Which arc do you think would be most suitable for the analysis of elements at trace levels?

The continuous DC arc will give a greater sensitivity, and will be better for trace analysis. This is because the higher current will make the anode hotter in the continuous DC arc, so that the sample will be atomised to a greater extent.

We can see that while the intermittent arc is better for precision analysis of major components, the continuous DC arc is better for trace analysis. Both types of discharge can be used in a single instrument.

These sources have been widely used in the past for qualitative and semi-quantitative analysis with spectrographic detectors. A prism is used to disperse the light onto a photographic plate. To obtain good resolution a very long path length was used, and the most obvious characteristic of these instruments is the enormous size of the spectrograph – a couple of metres long. The principle of the construction of the instrument is straightforward, and is illustrated in Fig. 7.7a:

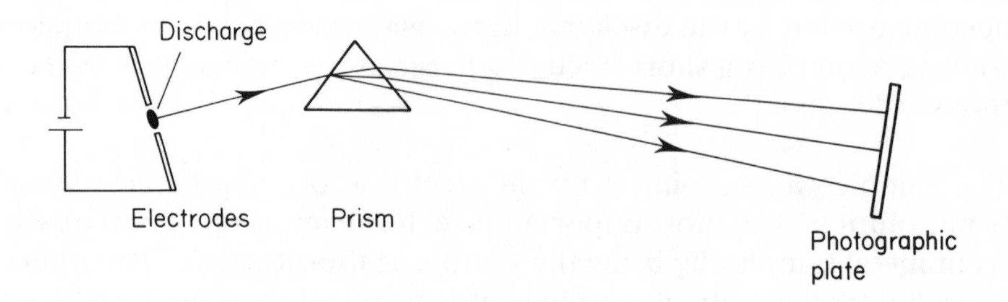

Fig. 7.7a. *The spectrograph*

The emission spectra of the elements present were recorded on the photograph as a series of vertical lines – a photograph of such a spectrum is shown at the front of the Unit. Comparison of the photograph with standard emission spectra allowed elements to be identified. An estimate of concentration could be made from the intensities of the lines.

More recent instruments use a polychromator with 20–60 photomultiplier detectors. Alternative sources such as the *glow discharge lamp* may also be incorporated into the instrument. In this lamp a low pressure argon discharge is used to vaporise the sample by cathodic sputtering, which is the same process used to to generate emission from a hollow cathode lamp. (Since the optics are essentially standard for any emission method, this type of spectrometer can also be used with an optional ICP source.)

These instruments are called *direct reading spectrometers* and are mainly used in the iron, steel and aluminium industries for the rapid analysis of metals and alloys. For these applications it is important to be able to monitor the non-metals carbon, phosphorus and sulphur which emit well below 200 nm. *Vacuum monochromators* which transmit light down to 165 nm must be used to allow these elements to be determined in metals.

SUMMARY AND OBJECTIVES

Summary

There are two main types of emission technique - flame and plasma. All emission techniques have the potential for simultaneous or rapid sequential determination of many elements. Spectral interferences are a major problem in all emission methods.

Flame emission methods, although more sensitive at longer wavelengths, are not as widely used as flame absorption methods. Two main flame emission techniques are in use – the simple flame photometer for the analysis of the alkali metals, and the modified atomic absorption spectrometer. In addition to the susceptibility to chemical and ionisation interferences also found in flame absorption methods, background emission, self-absorption and spectral interferences are common problems with flame emission methods.

Although plasma techniques have been known for some time, they have now been developed to the point where they are being increasingly used.

The ICP is the most popular plasma technique, combining high sample throughput for complex samples with the near freedom from chemical interferences, and detection limits which are comparable to flame-AAS in most cases and are superior to flame-AAS for the very refractory elements. On the other hand the ICP instruments are more expensive than flame-AAS instruments, and better precision is obtained with flame-AAS.

The other plasma techniques, the DCP and the MIP are less widely used as they are more susceptible to interferences and give poorer detection limits than ICP instruments. The MIP can be a very useful detector for gas chromatography.

Other emission methods which are useful for the direct analysis of solid samples include the DC arc and AC spark techniques. Spectral interferences present the main problem for all emission methods.

Objectives

You should now be able to:

- explain why emission spectroscopy is more sensitive at long wavelengths and higher temperatures (revision);
- appreciate why spectral interferences are more likely for emission methods compared to absorption methods;
- describe how spectral interferences may be overcome for both flame and plasma emission methods;
- describe the main components of a flame photometer;
- explain why flame-AES calibration plots are strongly curved, while ICP calibration plots are linear;
- explain how background emission may be corrected in flame-AES;
- compare the ways in which direct current (DCP), radiofrequency (ICP) and microwave (MIP) plasmas are generated, and compare the relative merits of these plasmas as emission sources;
- explain why the MIP can be used as a gas chromatography detector;
- describe how samples may be introduced into a plasma spectrometer;
- describe the characteristics of sequential and direct reading emission spectrometers;
- make reasonable estimates of sample throughput for different atomic spectroscopy techniques.

8. Perspectives – Present and Future

Overview

In this Unit we have studied a very wide, possibly bewildering, variety of atomic spectroscopic methods. It is important to try and get a sense of perspective about the various methods. Most analysts would be only too pleased to have a single instrument which would very rapidly determine a number of samples for as many elements as necessary at high sensitivity *and* precision. The instrument should also accept both solids and liquids with the minimum of sample pretreatment, carry out the analysis and produce a neat report containing results in which one can have complete confidence, with minimum operator effort. One last thing – it should be cheap to buy and cost little to run.

Basically what we would like to have is a black box where you pop the sample, untreated, in at one end and all the necessary information about the elements present is instantly printed out at the other end.

Well you probably realise that such an instrument does not yet exist, and given the last requirement, probably never will. Since we cannot have a perfect instrument we should ask ourselves which are our most important requirements and which type of instrument can best meet our needs. We may also want to ask whether one type of instrument is sufficient for our particular needs.

You probably also realise, or at least suspect, that none of the existing instruments can be considered the best technique for all applications. However, analytical atomic spectroscopy is a rapidly developing field, and great efforts are being made to overcome the limitations of many of the techniques. The specifications such as precision attainable, detection limits and multi-element capacity of the different methods appear to be converging.

In the first Section of this Part, we shall take a comparative look at the more commonly used current atomic spectroscopic methods. In the second Section we shall take a brief look at some of the ways in which these methods are being developed.

8.1. PRESENT METHODS

The requirements of different laboratories will vary so we can only discuss general requirements. We will also, for the sake of argument, ignore the fact that your laboratory may already have an instrument which, although perhaps not the most attractive method available, can adequately carry out the analyses required. There are several points which need to be considered including costs, detection limits, precision, versatility, speed of analysis, freedom from interferences and ability to analyse difficult elements.

The first question you should consider is not which atomic spectroscopy method, but is atomic spectroscopy the best technique? There are a wide variety of other methods available including:

classical analysis;

molecular spectroscopy;

electro-analytical techniques (ion selective electrodes, polarography, anodic stripping voltammetry etc);

chromatography (ion chromatography for solutions of metal ions, gas chromatography for organometallics);

X-ray fluorescence and X-ray diffraction;

mass spectrometry, and

neutron activation analysis.

The details of these methods are beyond the scope of this Unit. Many of these methods are described in other Units in the ACOL series, and the important point to bear in mind is that one of these methods may be more suitable than atomic spectroscopy for a given analysis. It is probably fair to say that few of these methods have the versatility and sensitivity of atomic spectroscopy for elemental analysis.

8.1.1. Costs

Assuming that we are intending to use atomic spectroscopy, one of the first considerations is cost. The most popular methods, excluding specialised equipment such as flame photometers and direct reading spectrometers, are flame-AAS, furnace-AAS, and ICP (sequential and simultaneous). The relative purchase costs are, with some overlap, in the sequence:

flame-AAS	$<$	furnace-AAS	$<$	sequential ICP	$<$	simultaneous ICP

Furnace-AAS is normally purchased as an attachment to a flame-AAS spectrometer. Hydride generation, though cheaper than furnace-AAS since a temperature programmer is not needed, can also be purchased as an attachment to be used either with a flame or with an ICP spectrometer. Simultaneous ICP is currently the most expensive method because of the large number of photomultipliers needed.

The current trends are for ICP instruments to become cheaper as the technology matures, and for more sophisticated and expensive flame-AAS instruments to be developed, so that the price differential between the most expensive flame-AAS and the cheapest ICP methods is tending to diminish.

The running costs can be an important consideration. Nitrous oxide-acetylene flames are more expensive to use than air-acetylene flames. The graphite tubes have to be regularly replaced in furnace-AAS. The most expensive running costs are however usually associated with the ICP which consumes large amounts of argon – 10 or 20 $dm^3\ min^{-1}$. Most of this gas is required as coolant to flow around the outside of the plasma, and this flow can be reduced by using a low power plasma or by using nitrogen as the coolant gas. Working at lower power can improve detection limits for some elements, but chemical interferences can become significant.

8.1.2. Detection limits and precision

Quoting detection limits is rather like standing on quicksand – the ground keeps shifting underfoot. Some comparative limits are shown in Fig. 8.1, but these should only be considered to be approximate, that is to a factor of two or three, as they vary from one type of instrument to another and improvements are being made all the time.

Element	ICP	DCP	Flame -AAS	Furnace -AAS	Hydride -AAS
As	15	30	100	1.0	0.01
Ba	0.2	–	8	0.2	–
Cd	1	5	1	0.01	–
Li	2	–	0.5	1.0	–
Mo	5	40	10	0.06	–
Se	15	50	100	0.2	0.01
Zn	1	5	0.6	0.005	–

Fig. 8.1. *Some detection limits in ng* cm^{-3} *(ppb). (0.02* cm^3 *solution for furnace-AAS, 50* cm^3 *solution for hydride-AAS assumed)*

Π Which is the most sensitive method for the determination of each of the elements in Fig. 8.1?

Hydride-AAS is the most sensitive method for As and Se.

Flame-AAS is the most sensitive method for Li, but there is little to choose between this method and ICP or furnace-AAS.

ICP and furnace-AAS have about the same sensitivity for Ba.

Furnace-AAS is the most sensitive method for Cd, Mo and Zn.

Furnace-AAS is by far the most sensitive method for most elements, while for a few elements such as As and Se the hydride generation method has the higher sensitivity. Furnace-AAS is therefore an essential method for any laboratory involved in the determination of trace metals.

Although the list of elements in the table is small it is possible to see that the techniques of flame-AAS and ICP are complementary. The more volatile metals such as Zn have slightly better detection limits for flame-AAS, while the refractory metals like Ba and Mo have better detection limits for ICP.

Furnace-AAS while being the most sensitive method, suffers from the poorest precision, about 5% for manual injection. Although this can be reduced by automatic sample introduction methods to about 1%, better precision can be obtained with ICP (0.5 to 2%) and especially flame-AAS (0.2 to 0.5%). This is one of the reasons why furnace-AAS is not normally used for measurement of higher concentration levels.

An important feature of the ICP method is that the precision is maintained over a much wider range of concentration because of the strongly linear calibration curve. This is an important point in automated analyses where a wide range of sample concentrations may be determined. While precision is poorer at low concentration for all methods, flame-AAS and furnace-AAS methods also suffer from loss of precision at high concentrations due to the flattening of the calibration curve.

8.1.3. Speed of analysis

The range of sample throughput for the various methods can vary widely.

Π Which of the three methods – ICP, furnace-AAS and flame-AAS, has the lowest sample throughput, and which has the highest sample throughput?

Furnace-AAS has the lowest sample throughput, because of the long cycle time involved in temperature programming.

ICP has the highest sample throughput for multi-element analysis because the elements can be determined simultaneously or by rapid sequential scanning. When only a few elements are to be determined, then flame-AAS can have a higher sample throughput because of the longer equilibration time needed in the plasma method. Automated flame-AAS spectrometers are available which can give very high sample throughput, measuring up to 8 elements or so at around 500 analyses per hour.

For non-routine analyses, the method development time, and sample preparation time should also be considered. For the ICP method development usually takes longer, but sample preparation may be faster since there is less need for sample pre-treatment to minimise chemical interferences.

8.1.4. Susceptibility to interferences and background

No technique is completely free from problems. We have seen that while both ionisation and chemical (or matrix) interferences are important in flame-AAS and furnace-AAS, ICP suffers more from spectral interferences. All these interferences can be dealt with, but care is needed to avoid making erroneous analyses.

For all methods background correction may be needed. Background correction is particularly important and virtually essential for furnace-AAS analysis at wavelengths below about 300 nm, because of the high molecular absorption and light scattering at these wavelengths. For all methods the best background correction methods determine the background both as simultaneously as possible and at as close a wavelength as possible to the sample measurement. Methods such as the hydride generation technique which involve separation of the sample before analysis are less susceptible to background problems, but chemical interferences can still occur.

8.1.5. Direct analysis of solids

Π Which of the three methods – flame-AAS, furnace-AAS and ICP – can be used for the analysis of solids without dissolution of the sample?

All three techniques can be used for the analysis of solids, with varying degrees of ease.

Furnace-AAS is the technique for which direct analysis of solids is relatively straightforward, (not counting the specialised direct reading spectrometers), since solid samples can be placed directly into the furnace tube. Since the samples are very small homogeneity can sometimes be a problem.

Slurry injection can be used with both flame and plasma methods. Direct analysis of solids is also possible with ICP by inserting the sample on the end of a graphite rod.

8.1.6. Which technique?

There is often lively debate about which is the best atomic spectroscopic technique. Unfortunately there is no simple answer. It all depends on the type of analysis to be carried out and even then it may be desirable to have more than one technique available.

It is probably fair to say that the flame-AAS method is the most necessary method to have in a laboratory. For analyses where only a few elements are likely to be determined at moderate or high concentrations then flame-AAS is probably the best method. For the "one-off" or occasional analysis, method development is easier for flame-AAS, and needs less operator skill than the ICP. If complex samples are to analysed for a wide variety of elements including those which are difficult in flame spectroscopy such as boron and phosphorous, then ICP is the preferred method. Typical applications for ICP include the multi-element analysis of complex geochemical samples, and river water samples. Even with an ICP spectrometer, it is useful to have a flame-AAS spectrometer for checking the ICP analysis (to avoid using lines with spectral interferences).

No other technique is as useful as furnace-AAS for trace element determination in general, although there are several useful specialised methods for certain elements such as mercury and the volatile hydride-forming elements. Since furnace and the other methods are available as flame-AAS accessories, this supports the argument for having flame-AAS.

SAQ 8.1a

Which is the best method for the following analyses?

(*i*) Aqueous samples of selenium at levels of 10 ng cm^{-3} (ppb).

(*ii*) River water samples for 20 or 30 elements at levels from 0.01 to 100 mg dm^{-3} (ppm).

(*iii*) Sodium in body fluids.

(*iv*) Iron in an organometallic compound (10%).

(*v*) Lead in street dust (0.5%).

(*vi*) Multi-element alloy determination, including phosphorus.

(*vii*) Trace levels of cadmium in food ($<$ 0.1 mg kg^{-1}).

(*viii*) Mercury in fish ($<$ 0.1 mg kg^{-1})

8.2. FUTURE DEVELOPMENTS

Many of the improvements in atomic spectroscopy have come about as a spin-off of the dramatic developments in computing in recent years, which have helped to improve the 'user-friendliness' of the instruments, and to increase the precision of measurement and the sample throughput. It is no accident that the growth of interest in plasma emission spectroscopy has followed the developments in computing, since simultaneous or rapid scanning multi-element analysis with emission methods is an obvious application for computer control. All methods are capable of being automated, again benefitting from the use of cheap but powerful computers.

Flow injection methods, in which solution samples separated by solvent are flowed continuously through narrow-bore tubing into a nebuliser for ICP analysis are available to increase the rate of sample throughput.

Although software and automation developments will continue, along with the integration of atomic spectrometers into laboratory information management systems (LIMS), these developments are not fundamental ones.

In this Section we shall concentrate on some of the current instrumental developments which are intended to overcome some of the limitations of the existing commercial instruments. We shall consider each of the major techniques – flame, furnace and plasma in turn, and we shall also consider a few of the developments in hyphenated methods (ie the interfacing of other techniques to atomic spectrometers.)

8.2.1. Flame spectroscopy

There seems to be little that can be done to overcome the basic limitations of flames. Higher flame temperatures cannot be reached to overcome chemical interferences. Residence times can be improved to some extent to enhance sensitivity, as we saw in Part 5, by means of various atom trapping techniques.

The most likely developments are in the light sources used for flame spectroscopy. Some improvement can be made as we have seen by using computer control of several hollow cathode or electrodeless discharge lamps. A much more fundamental development is to replace the atomic resonance lamp with a continuum light source such as a high pressure xenon arc lamp which emits radiation continuously across the entire uv-visible spectrum. At first sight this might seem strange. After all, we have already seen that the rise of AAS as an important analytical tool was largely due to the development of the hollow cathode lamp by Walsh to utilise the 'lock and key' effect.

The answer to this apparent contradiction lies in the improvements that have been made in monochromator resolution. The monochromators available in the 1950s did not have sufficient resolution to only transmit the wavelengths at which the analyte atoms absorbed. This is illustrated in Fig. 8.2a, which shows an atomic absorption line of bandwidth around 0.005 nm as part of the light transmitted by a monochromator of bandpass 0.1 nm, when a continuous source is used. Nearly all the light reaching the detector will be from wavelengths other than the absorption line. As a result the detector signal will change only slightly with the introduction of the sample, and the sensitivity of the technique will be extremely poor (see also Section 3.2).

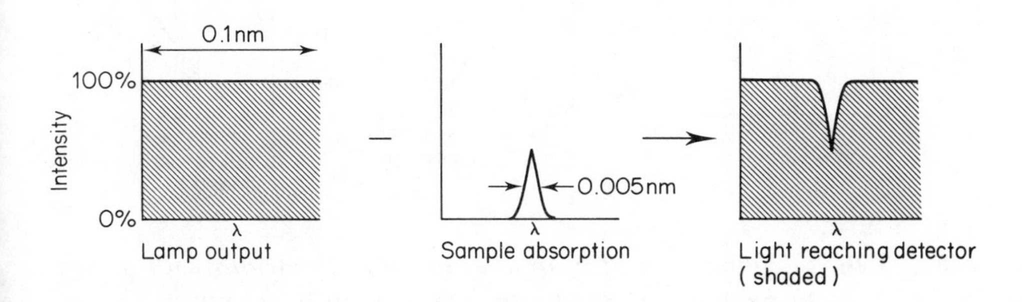

Fig. 8.2a *Comparison of atomic absorption profile with output of a high pressure xenon lamp for a monochromator bandpass of 0.1 nm*

The situation changes if we can use a monochromator with a very narrow spectral bandpass of about 0.002 nm, less than the width of the absorption line.

Π Which type of monochromator would you use to obtain a spectral bandpass of 0.002 nm?

(*i*) Conventional grating.

(*ii*) Echelle grating.

(*iii*) Prism.

The correct answer is (*ii*).

The echelle grating is the only type of monchromator that can conveniently be used for this narrow bandpass. Conventional gratings are often used to give bandpasses down to 0.01 nm or so, but the extra resolution of a factor of about ten for echelle gratings is necessary for monochromators of 0.002 nm bandpass.

The effect of the higher resolution can be seen in Fig. 8.2b.

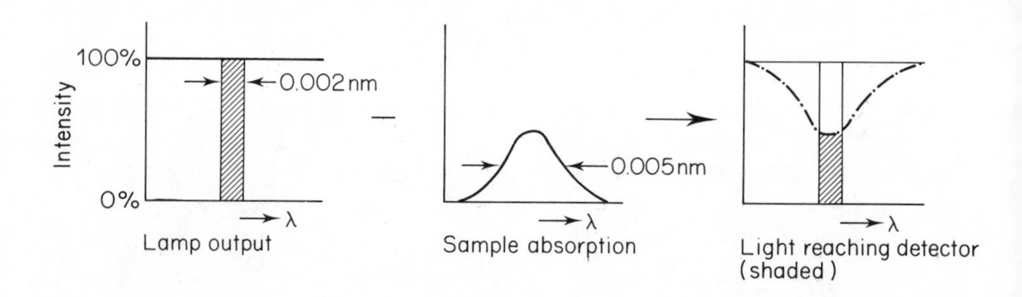

Fig. 8.2b *Comparison of atomic absorption linewidth with a very narrow monochromator bandpass of 0.002 nm*

The light reaching the detector is shaded in Fig. 8.2b as in Fig. 8.2a. The important point to note is that with the broad bandpass monochromator, the sample absorption makes very little difference to the light intensity reaching the detector, but for the narrow bandpass monochromator, the same amount of sample absorption strongly reduces the light reaching the detector.

The net effect of using a narrow bandpass monochromator is to greatly increase the sensitivity of the technique, known as continuum source AAS or CSAAS.

The use of a continuum source has two advantages over conventional or line source AAS.

(*i*) Spectra can be scanned. The method is therefore inherently a multi-element one. Only one souce is needed and any element can, in principle, be determined.

(*ii*) The dynamic range is much greater than for line source AAS. For higher concentrations, which normally give a flattening of the calibration curve, the absorption measurements can be made on the *side* of the absorption line, giving a linear response with lower sensitivity. For lower concentrations the *centre* of the absorption line is used.

You might think it rather surprising that a single light source can be intense enough to be usable for all elements. In fact the output of a xenon arc is intense enough for the detection limits obtained with CSAAS to be comparable with those obtained with line source AAS for wavelengths above 280 nm. Unfortunately the uv output of the lamps falls off rapidly below 280 nm and poorer detection limits are obtained. It is not easy to obtain high intensity far-uv output with any continuous output lamp, but lasers offer a potential solution to the problem. Lasers give very intense output of very narrow bandwidth, normally at a fixed wavelength. The development of the *tunable* uv laser in which the output wavelength can be varied should allow scanning of the uv absorption spectrum.

8.2.2. Furnace spectroscopy

Graphite furnace methods will continue to be the most sensitive in atomic spectroscopy, but attempts are being made to overcome some of the limitations of the methods. The use of totally pyrolytic graphite tubes, and platform methods to reduce interference effects are rapidly becoming standard, as is the use of deuterium arc, Zeeman or Smith-Hiefte background correction procedures.

One of the main problems with furnace-AAS is the long cycle time, with two or three minutes required for each measurement. The sample throughput can be increased by using the furnace as a multi-element method. This can be done either as an absorption method or as an emission method.

(*i*) Continuum source AAS can be used with a graphite furnace to generate the atom cell just as easily as it can be used with a flame atom cell. The cycle time of 2–3 minutes can be used to determine several elements. There is some loss of sensitivity compared to line source AAS.

(*ii*) Any emission source is inherently multi-element. The main problem with the graphite furnace is the low temperature of operation. Thus it is a suitable source for easily excited elements emitting at long wavelengths, but less suitable for elements emitting at short wavelengths. One possible way round this problem is to pass the atoms from a furnace into a low pressure electrical discharge to generate excited atoms.

Π Would you expect graphite furnace-AAS to be more sensitive than graphite furnace-AES for:

(*i*) Pb (283.3 nm), and

(*ii*) Li (670.8 nm)?

(*i*) Furnace-AAS would be the best technique for Pb analysis. At this short wavelength the emission intensity will be rather low.

(*ii*) Furnace-AES would be the most sensitive method for Li analysis. Lithium is an easily excited element, and the emission at 670.8 nm will be relatively intense. The detection limits for Li for furnace-AES are 5 to 10 times better than for flame-AAS.

8.2.3. Plasma spectroscopy

There is considerable interest in improving existing plasma methods, and most interest seems to be in developing the ICP plasmas because of the superior detection limits and relative freedom from interferences. Low power ICP torches to reduce running costs are being developed. Attempts to improve nebuliser design, which should increase analytical precision, are being made. The direct analysis of solid samples is still some way from being adopted as a routine method. Other methods of gas or liquid sample introduction are also being developed. Hydride generation can be used with ICP analysis, and the ICP can be used as an hplc (high performance liquid chromatography) detector.

The ICP seems especially prone to the development of 'hyphenated' or hybrid techniques, resulting in an explosion of new acronyms. Examples are ICP-MS (ICP interfaced to a mass spectrometer), ICP-AFS (ICP as an excitation source for atomic fluorescence spectroscopy), ICP-FTS (ICP fourier transform spectroscopy). The basic ICP technique is now often given a longer acronym to distinguish it from the various 'hyphenated' methods – both ICP-AES (ICP atomic emission spectroscopy) and ICP-OES (ICP optical emission spectroscopy) are used.

ICP-MS

The idea of the ICP-MS technique is to use the ICP as an *ion source* for a mass spectrometer. A mass spectrometer can separate ions and identify them according to their masses.

The method eliminates spectral interferences which are such a problem in ICP-AES. Detection limits are about a factor of ten better than ICP-AES because ions can be detected at low levels with greater precision than photons.

The interface between the ICP and the MS is important since the former is at atmospheric pressure while the latter is under a high vacuum. Two nickel cones or skimmers with a small (1 mm) hole are used to transmit the ions from the plasma, with a progressive reduction in pressure between the cones. The interface is described in Fig. 8.2c.

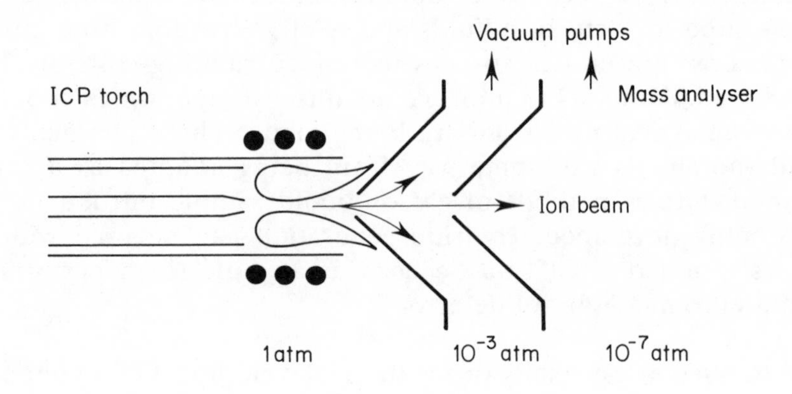

Fig. 8.2c *The ICP-MS interface*

Π Which part of the plasma do you think would be the best ion source – the tail plume (6000 K), or the hotter regions of the plasma (7500 K)?

The hotter regions of the plasma should be used, since the concentration of ions is greater in the hotter regions. Remember that even more energy is required to ionise an atom than to populate atomic excited states.

The levels of ionisation in the plasma are very high – at 7500 K only the inert gases and the atoms H, C, N, O, F, Cl and Br are less than 10% ionised. Most elements are more than 90% ionised at this temperature.

The main problem with ICP-MS is that one type of interference (spectral) is replaced by another. Background *molecular* ions, such as Ar_2^+, O_2^+ and N_2^+ are usually present along with atomic Ar^+ ions. These ions will be detected at the same mass numbers as some elements. For example the atomic argon ions at mass 40 amu will overlap and interfere with calcium ions which also have a mass of 40.

Π Can you work out which elements have the same mass as the molecular ions of oxygen (28 amu) and nitrogen (32 amu)?

Silicon will give ions of mass 28 amu, while sulphur will give ions of mass 32 amu. Therefore problems can be expected in the analysis of these two elements by ICP-MS.

These mass interferences are usually confined to low atomic mass elements, but are more of a problem if organic solvents are used. This is because several higher mass ions can be formed from the fragmentation of the solvent molecule. The interference can often be side-stepped as most elements give more than one isotope that can be analysed, and some isotopes will be less prone to interference.

The technique of ICP-MS is still in its infancy and considerable improvements may be expected.

ICP-AFS

Another way of overcoming spectral interferences in ICP spectroscopy is to use the ICP as a *light source* for multi-element atomic fluorescence spectroscopy. This is illustrated in Fig. 8.2d

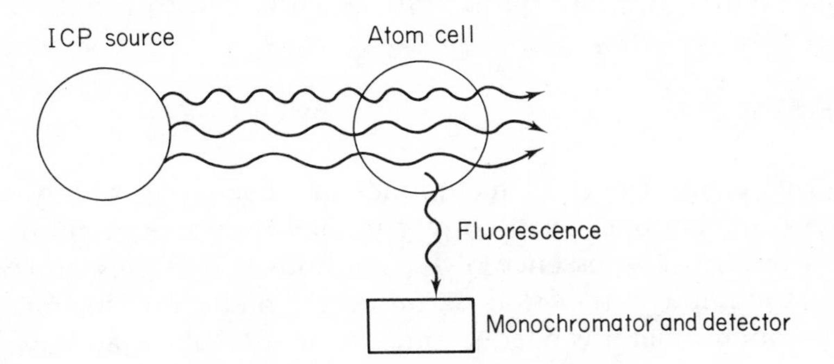

Fig. 8.2d *The ICP-AFS technique*

The difficulty with multi-element fluorescence spectroscopy is that of obtaining a sufficiently intense light source to excite fluorescence from several elements. Continuum sources are too weak and suitable lasers are not yet available. The ICP can be thought of as an intense multi-element line source.

Π Which of the following will give rise to fluorescence from ground state atoms in the atom cell?

(*i*) Resonance emission lines from the light source.

(*ii*) Non-resonance emission lines from the light source.

Only resonance emission lines will give rise to fluorescence, as we saw in Part 1 of the Unit.

Note that although light of several different wavelengths is emitted by the source, only a few wavelengths are emitted as fluorescence. These correspond to the resonance lines of atomic species which are present in both the source and the atom cell.

Although the ICP emission spectrum is very complex, the fluorescence spectrum is rather simple and generally free from spectral interferences.

A low power ICP can be used as an atom cell to generate mainly ground state atoms.

ICP-FTS

Fourier transform (FT) techniques are becoming widely used in other branches of spectroscopy, and may yet prove useful in atomic spectroscopy. The essence of FT methods is that they are *parallel-access* methods rather than *serial-access* methods. This means that no monochromator is needed and information about all wavelengths in the spectrum is obtained simultaneously. In this, it is more like using a polychromator than a scanning monochromator (one wavelength at a time), except that information about the *complete* spectrum is obtained continuously. How FT methods work is too complex to go into here, except to say that an interferometer is used to generate an interference pattern from all the different wavelengths of light present. The important point is that the methods are very sensitive and capable of high resolution. For example a resolution of 1 part in 300 000 can be obtained.

Π What resolution can be obtained using FT methods for a spectral line at 300 nm:

(*i*) 0.1 nm

(*ii*) 0.01 nm

(*iii*) 0.001 nm

(*iv*) 0.0001 nm.

The correct answer is (*iii*), since 0.001 nm in 300 nm corresponds to 1 part in 300,000.

So we can see that FT methods promise to be competitive with echelle grating monochromators where high resolution is needed.

One of the limitations of FT techniques is the mechanical tolerances which must be reached in the interferometer, which uses an accurately controlled moving mirror. The tolerance to which the position of the moving mirror must be known must be less than the wavelength of the light being studied.

Π Will the mechanical tolerances allowed in a FT spectrometer be smallest for:

(*i*) the infra-red;

(*ii*) the visible region;

(*iii*) the ultra-violet.

The correct answer is (*iii*).

Ultra-violet light (200–400 nm) wavelengths are shorter than for visible or infra-red light.

The high mechanical precision needed is one reason why FT uv-visible instruments have not been so widely developed as FT infra-red instruments, but FT-AES instruments have been described and shown to be promising.

The high resolution and the ability to obtain a complete spectrum may turn out to make FT-AES a useful instrument for non-routine analyses.

SUMMARY AND OBJECTIVES

Summary

There are a great many techniques available for analytical atomic spectroscopy. In many respects these techniques are complementary, since no single technique is suitable for all analyses. There are several specialised techniques for particular applications, such as hydride generators, flame photometers and direct reading spectrometers. For general purpose analyses, however there are three widely used techniques.

Flame-AAS is the most accurate method, for most elements at moderate concentrations. ICP is suitable for the multi-element analysis of complex samples, and is less prone to chemical interferences than other methods. Furnace-AAS is the most suitable technique for trace analysis, although the precision is rather poor, and the sample throughput is low.

All methods suffer from disadvantages, and the effort currently being expended to overcome these is an indication of the popularity of analytical atomic spectroscopy. Many of the developments are due to the more general use of computing for instrument control and data handling. Other developments involve a more fundamental reassessment of techniques. Continuum source AAS shows potential as a multi-element technique, both with flame and furnace atom cells. Furnace emission methods also show some potential. New ICP torches and sampling devices are under development. The ICP is also being developed as an ion source for mass spectrometry, a light source for atomic fluorescence spectroscopy, and a chromatography detector. Fourier transform methods also show promise.

Atomic spectroscopy has been an essential part of the analysts armoury for several decades, and the current state of the art (and it is still sometimes an art as well as a science) suggests that analytical atomic spectroscopic techniques will continue to be a vital and important subject.

Objectives

You should now be able to:

- identify the most appropriate technique for a given type of analysis;
- list the main advantages and disadvantages of the most widely used atomic spectroscopic methods;
- describe some of the ways in which new developments are attempting to overcome these limitations;
- outline the basic principles of 'hyphenated' methods such as ICP-MS or ICP-AFS.

Appendix

Precision, accuracy and detection limits

Throughout this Unit we use the terms *precision* and *detection limit*. More rarely the term *accuracy* is used. These, and other terms used to express our confidence in our results and the sensitivity of various techniques, are described in this appendix, so that they can be referred to as necessary while you work through the Unit. The discussion of *standard deviation* involves some statistics, but mathematics is kept to a minimum.

Precision is a measure of how repeatable a given set of results are, or how close they are to each other. For example, when we use a burette for a titration, the results of duplicate titrations should agree to 0.1 cm^3 or better. The burette is said to have a precision of ± 0.05 cm^3. If we were to measure the length of a piece of A4 paper several times with a simple ruler, we should expect the results all to agree to within perhaps 0.5 mm. In this case the precision of the measurement would be ± 0.25 mm.

Precision then is a measure of our confidence in the measured quantity. It is important to realise that precision is not the same as accuracy. Accuracy is a measure of the closeness of the result to the correct value and is much more difficult to obtain than precision. For example the burette or the ruler in the above examples may be incorrectly calibrated, leading to consistent errors in the measurements.

A recent example highlights the problems of achieving accuracy in atomic spectroscopy, which can be particularly severe for the determination of trace elements in complex samples. The data being produced for lead and cadmium levels in foodstuffs by various analysts were found to be significantly in error, in many cases. (For a more detailed account see the article in *Chemistry in Britain* by J. C. Sherlock et al, (Volume 21, page 1019, (1985)), and a regular stream of letters discussing the controversy in Chemistry in Britain over the following months). The errors appear to be due to inability to achieve adequate blanks or use appropriate standards, or lack of appreciation of the detection limits of the methods used.

There are two points to bear in mind, firstly we should have a clear understanding of how precise our analysis is, and secondly we should try to understand the technique well enough to be aware of the possible sources of error.

There is another useful rule to bear in mind in any form of analysis. The greater the precision needed the greater the time needed. Accuracy takes even longer to achieve. Since time is often money, there is little point in achieving more precision than is necessary. In many analyses for trace elements it is simply necessary to know whether a sample is present above or below a given threshold level, and there is no point in striving for a precision of $\pm 0.5\%$ in the measurement when $\pm 10\%$ is adequate.

We can categorise errors into two types.

(*i*) Random errors – the sort of error which leads to different results when the measurement is repeated, eg the error in end-point determination in a titration.

(*ii*) Systematic errors – one-way errors which can arise through incorrect calibration or errors of method eg all forms of atomic spectroscopy are prone to the presence of *interferences* which give a reading which is consistently low or high.

Π Which type of error – random or systematic – can be reduced by taking the average of several readings?

Random errors can be reduced by averaging, since some readings will be on the low side, while others will be on the high side, and the extremes will tend to cancel when we average.

Systematic errors cannot be changed by averaging. If an interfering process is causing a low reading then no amount of repetition will make the reading any higher.

When we use the term precision we are using a measure of the random errors in an analysis. Systematic errors are more difficult to account for and are responsible for any significant differences between the measured value and the correct value.

We can put the term precision on a more mathematical basis as follows:

The average of a number of readings is called the *mean*, $\bar{x}$.

The *standard deviation*, s, is a measure of the spread of results about the mean.

If we take a series of n readings, such as absorbance measurements, A_1, A_2, A_3, ... A_n, then the mean or average value is:

$$M = \frac{(A_1 + A_2 + \ldots\ldots A_n)}{n}$$

The standard deviation is defined in terms of the individual deviations of each result from the mean, $D_1 = A_1 - \bar{x}$ and so on.

$$s = \sqrt{\frac{(D_1^2 + D_2^2 + \ldots . D_n^2)}{n - 1}}$$

For example, say we have a series of eleven absorbance readings of 0.500, 0.490, 0.500, 0.520, 0.500, 0.480, 0.500, 0.500, 0.500, 0.510, and 0.500, we can add all these values up and divide by 10 to get the mean of 0.500. Using the formula for s:

$$s = \sqrt{\frac{(0 + 0.0001 + 0 + 0.0004 + 0 + 0.0004 + 0 + 0 + 0 + 0.0001 + 0)}{10}}$$

$= 0.010$

So we can say that the average value is 0.500 with a standard deviation of 0.010. (You may find this calculation easier with a pocket scientific calculator, most of which have mean and standard deviation functions. It often more convenient to express the result as the percentage coefficient of variation (% C.V.), which is simply the standard deviation expressed as a percentage of the mean value. In the above case:

$$\% \text{ C.V.} = (0.010 / 0.500) \times 100$$

$$= 2.0$$

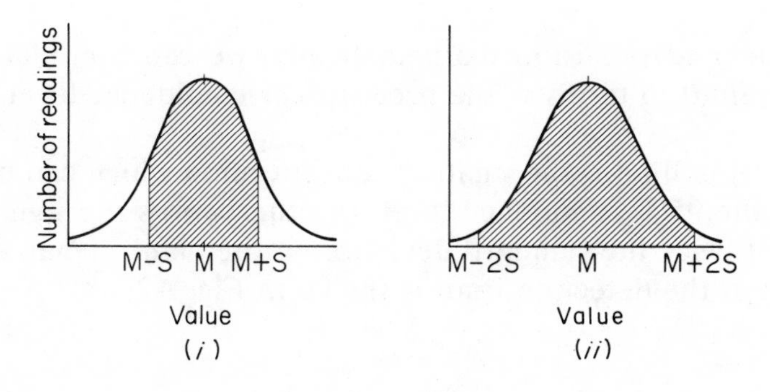

Fig. A1. *Error curve with confidence limits (i) ± s and (ii) ± 2s*

The physical significance of the standard deviation is that if a large number of measurements is made, then 68% of the results will differ from the mean value by no more than s. This is shown as the shaded area in Fig. A1(*i*), where the curve represents the distribution of results. In practice we usually desire to be a little more confident about our results than this because it means that about one result in every three will lie outside the stated precision. It is more usual to use twice the standard deviation as a measure of precision, since 95% of results lie between $\bar{x} - 2s$ and $\bar{x} + 2s$, the shaded area in Fig. A1(*ii*). In this case only one result in twenty is likely to lie outside the stated precision.

Π In the above example, is the precision of the measurement, expressed as a percentage:

(*i*) 1%

(*ii*) 2%

(*iii*) 4% ?

The correct answer is (*iii*).

The precision is twice the standard deviation , or in percentage terms, twice the C.V.

This definition of precision is sometimes referred to as the 95% confidence level.

Having defined precision mathematically, we can now define the *detection limit* in terms of the precision or confidence level.

The detection limit is the smallest concentration which can be measured at the 95% confidence level. In other words the signal must be at least twice the standard deviation in the blank signal. A measurement at the detection limit is shown in Fig. A2.

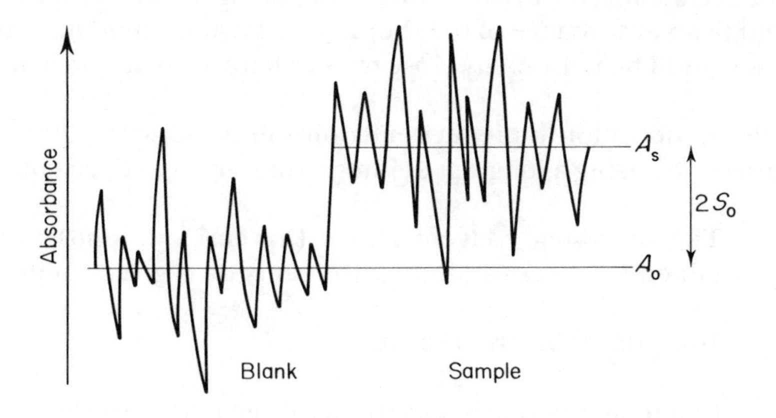

Fig. A2. *The detection limit*

Mathematically:

$$A_S = A_0 + 2 \times s_0$$

where A_S is the sample absorbance, and A_0 and s_0 are the blank signal and the standard deviation in the blank signal.

Another term indirectly related to precision which is sometimes used is the *characteristic concentration*. It is defined as the concentration which gives rise to 1% absorption of the incident light in an AAS experiment. This is equivalent to an absorbance of 0.0044.

Π Which of the two terms precision and characteristic concentration depend on the noise level in the spectrometer?

The precision depends on the noise level. The characteristic concentration does not.

It should be realised that the precision varies from day to day and from one spectrometer to another. Therefore detection limits, which are based on the precision of the technique, can only be considered approximate, to a factor of at least 2.

Since the standard deviation in the blank signal will usually correspond to an absorbance of 0.001 or a little greater, then the detection limits should be reasonably close to the characteristic concentration.

Since the detection limits depends on the noise level, they can be improved by using a long signal integration time to reduce the noise.

Π The following concentrations (ng cm^{-3} or ppb) were obtained for a series of measurements on a given sample:

105, 106, 100, 109, 112, 104.

Calculate the mean and standard deviation for these results.

If the correct result for the analysis is 102 ng cm^{-3}, does the measured value agree within the precision of the measurement?

The mean value is 106 ng ml^{-1}.

The standard deviation is 4.15 ng cm^{-3} (ie about 4% C.V.).

The precision of the measurement is $2 \times s = \pm 8.3$ ng cm^{-3}. The measured range is therefore 98–114 ng cm^{-3}. The correct value, 102 ng cm^{-3}, falls within this range and is therefore within the precision of the measurement. It must be admitted that the precision in this case is rather poor.

Self Assessment Questions and Responses

SAQ 1.2a

> The characteristic yellow light emission of sodium, seen in flames containing sodium salts, or from sodium street lamps, occurs at a wavelength of 589 nm.
>
> For photons of this wavelength, calculate
>
> (*i*) the frequency;
>
> (*ii*) the wavenumber;
>
> (*iii*) the energy.

Response

This SAQ is worked through later in Section 1.2.1. The correct answers are:

(*i*) the frequency is 5.093×10^{14} Hz.

(*ii*) the wavenumber is 1.698×10^6 m^{-1} or 1.698×10^4 cm^{-1}.

(*iii*) the energy of the photon is 3.375×10^{-19} Joules. This may also be expressed as the energy of one mole of photons, a more familiar unit to chemists, and has the value 203 kJ mol^{-1}.

If you experienced any problems with this calculation, you should work through Section 1.2.

****

SAQ 1.2b

Arrange the following regions of the electromagnetic spectrum in order of (*a*) increasing wavelength, and (*b*) increasing energy:

(*i*) ultra-violet;

(*ii*) infra-red;

(*iii*) visible.

Response

(*a*) the order of increasing wavelength is:

ultra-violet < visible < infra-red

The visible region of the spectrum is from approximately 400 nm (blue) to 700 nm (red).

(*b*) the order of increasing energy is

infra-red < visible < ultra-violet

Since the energy is inversely related to the wavelength, as can be seen by combining Eq. 1.1 and 1.5 in Section 1.2 to yield:

$$E = hc / \lambda$$

****

SAQ 1.2c

Fig. 1.2a, (*i*), (*ii*) and (*iii*) depict the three energy level diagrams for the processes of absorption of radiation, emission of radiation and fluorescence. Identify each process.

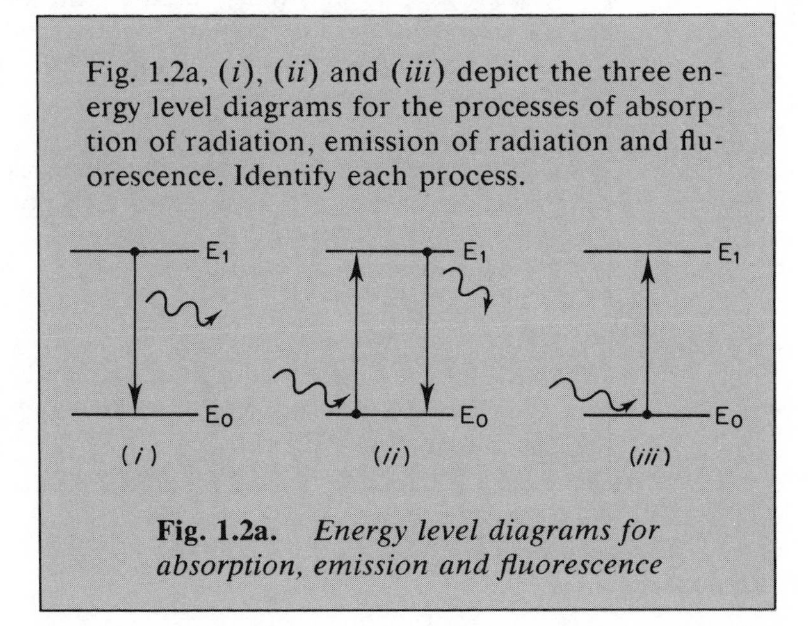

Fig. 1.2a. *Energy level diagrams for absorption, emission and fluorescence*

Response

Energy level diagram (*i*) shows the process of *emission*, in which the atom in an excited energy level, E_1, reverts to the ground state with the emission of a photon.

Energy level diagram (*ii*) shows the process of *fluorescence*, in which the atom absorbs and re-emits a photon.

Energy level diagram (*iii*) shows the process of *absorption*, in which an atom in its ground energy level, E_o, is promoted to a higher energy level by the absorption of a photon.

****

SAQ 2.2a

How sensitive are atomic absorption spectroscopy and atomic emission spectroscopy to variations in the temperature of the atom cell? Flames do flicker which shows that there may well be significant temperature fluctuations. To investigate this sensitivity calculate the effect of a 10 K variation in flame temperature on N_1/N_0, for sodium atoms, by calculating N_1/N_0 at 2000 K and at 2010 K.

Use the results of these calculation to estimate the effect of the 10 K temperature rise on

(*i*) I_e, and

(*ii*) I_a.

(For sodium atoms E = 203 000 J mol^{-1} and $g_1/g_0 = 2$.)

Response

A 10 K temperature rise will increase I_e by 6%, but have little effect on I_a.

Again, using the Boltzmann equation, at 2000 K where RT = 8.314 × 2000 = 16,628 J mol^{-1}:

$$N_1/N_0 = 2\exp(-203,000/16,628)$$

$$= 1.00 \times 10^{-5}$$

$$\text{At 2010 K where } RT = 8.314 \times 2010$$

$$= 16\ 711 \text{ J mol}^{-1}:$$

$$N_1 / N_0 = 2 \exp(-203,000/16 - 711)$$

$$= 1.06 \times 10^{-5}$$

There is then a 6% increase in N_1 which will give rise to a 6% increase in I_e (since I_e is proportional to N_1).

There is little change in N_0 (0.9999900 → 0.9999894) so there will be little change in I_a, which is proportional to N_0.

We can expect that temperature stability will be very important in AES, and that noisy signals may result from variations in flame temperatures.

****

SAQ 2.3a

The spectral line width of the 589.6 nm sodium line has been measured as 0.003 nm at 1000 K and 0.005 nm at 3000 K. Do you think this observation is consistent with the line width being determined by Doppler broadening?

Response

The observation is consistent the line width being determined by Doppler broadening.

As a quick check that we are on the right lines, you may notice that $\Delta\lambda_{1/2}$ is proportional to the square root of T, Eq. 2.3, and that as the ratio of the two given temperatures is 3:1 we would predict the line widths to be in the ratio $\sqrt{3:1}$ as is observed (0.005/0.003 = 1.6 $\approx \sqrt{3}$).

More specifically we can calculate $\Delta\lambda_{1/2}$ remembering that careful consideration must be give to the units involved. If we use m s^{-1} for the speed of light, the units of A_r must be consistent with this. Therefore SI units must be used, and in particular the relative atomic mass must be expressed as 0.023 kg mol^{-1} for sodium.

At 1000 K

$$\begin{aligned} \Delta\lambda_{1/2} &= (2/3 \times 10^8)(2 \times 8.314 \times 1000/0.023)^{1/2} \times \lambda \\ &= 5.6 \times 10^{-6} \times \lambda \\ &= 5.6 \times 10^{-6} \times 589.6 \text{ nm} \\ &= 0.0033 \text{ nm} \end{aligned}$$

which is close to the observed value of 0.003 nm.

Similarly at 3000 K

$$\begin{aligned} \Delta\lambda_{1/2} &= (2/3 \times 10^8)(2 \times 8.314 \times 3000/0.023)^{1/2} \times 589.6 \\ &= 0.0058 \text{ nm} \end{aligned}$$

which is again close to the experimental value.

We may conclude that the lines are Doppler broadened.

****

SAQ 2.3b

A hollow cathode lamp (see Part 3) used in atomic absorption spectroscopy emits atomic iron lines but also contains neon gas and emits atomic neon lines at nearby wavelengths. If the spectral lines are mainly Doppler broadened, which of the following statements is correct?

(*i*) The neon and iron line widths are about the same.

(*ii*) The iron lines are wider than the neon lines.

(*iii*) The neon lines are wider than the iron lines.

The relative atomic masses of neon and iron are 20 and 56 respectively.

Response

The correct response is (*iii*).

Since the lines are Doppler broadened we can apply Eq. 2.3

$$\Delta \lambda_{1/2} = (2/c)\,(2RT/A_r)^{1/2}\,\lambda$$

Since the two elements are at the same temperature and if the wavelengths are not too different:

$$\Delta \lambda_{1/2} \propto A_r^{-1/2}$$

So the higher the relative atomic mass the narrower the spectral lines. This is reasonable because heavier atoms move more slowly than lighter atoms.

In particular, neon will give broader spectral lines than iron.

****

SAQ 3.1a

The measured absorbance for a particular sample is 0.699. What percentage of light is transmitted by the sample?

Response

The correct answer is $I_t = 20\%$.

Since the absorbance is known and I_0 is 100%, the I_t can be calculated using Eq. 3.2:

$$A = \log (I_0/I_t)$$

$$\text{Substituting, } 0.699 = \log (100/I_t)$$

$$0.699 = \log 100 - \log I_t$$

$$\log I_t = \log 100 - 0.699$$

$$= 2.0 - 0.699$$

$$= 1.301$$

$$\text{taking antilogs, } I_t = 20\ \%$$

****

SAQ 3.1b

The *characteristic concentration*, discussed in the Appendix, is a useful measure of how sensitive a technique is. It is the concentration of an element which gives rise to *1% absorption* or 99% transmission of the incident radiation. What is the corresponding value of absorbance when $I_t = 99\%$?

Response

The correct answer is $A = 0.0044$.

Again Eq. 3.2 is used, but this time the unknown quantity is A. Since I_t is 99% and I_0 is 100%:

$$A = \log(100/99)$$

$$= 0.00436$$

Note that A has no units since it is the logarithm of a number.

****

SAQ 3.1c

The following values of A were obtained for a series of standard zinc solutions:

A	0.0	0.152	0.298	0.450	0.600	0.740	0.860	0.940
$[Zn]$ /mg dm^{-3}	0	2	4	6	8	10	12	15

Plot the calibration curve and determine the concentration of two unknown solutions with absorbances of 0.225 and 0.900 respectively.

Response

The concentrations of the unknowns are 3.0 ppm and 13.2 ppm.

The data are plotted on Fig. 3.1d below.

The concentrations are simply read off the plot by drawing a horizontal line from the measured absorbance reading to meet the curve. A vertical line from this point is then drawn to the concentration axis and the value of the concentration determined.

Note that the plot starts to curve at 12ppm and flattens off above 15 ppm, so that 0 to 15 ppm corresponds to the useful range of the curve.

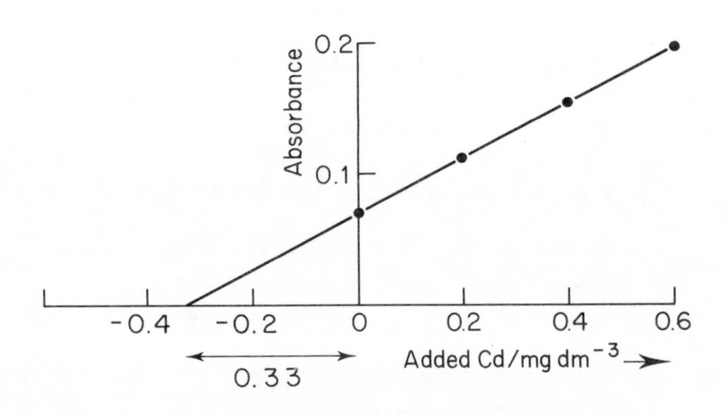

Fig. 3.1d. *Graph of absorbance against concentration*

****

SAQ 3.1d

1.20 g of an alloy containing nickel was dissolved in hydrochloric acid and the solution made up to 100 cm^3. This solution was found by AAS to contain 150 mg kg^{-1} of nickel. What is the percentage of nickel in the original sample?

Response

The correct answer is 1.25%.

If the solution contains 150 mg kg^{-1}, or 150 mg dm^{-3}, then 1 dm^3 of solution would contain 150 mg of nickel. A solution volume of 100 cm^3 would therefore contain 15 mg of nickel.

Since the whole of the original 1.20 g sample was dissolved in this solution, we can say that 1.20 g of sample contains 0.015 g of nickel. Hence:

$$\text{Percentage of Ni} = 100 \times (0.015/1.20)$$

$$= 1.25\%$$

****

SAQ 3.2a

> The linewidth of the 589 nm sodium line is about 0.005 nm at flame temperatures of 2000 K. Would you expect the linewidth of the resonance emission line from a hollow cathode sodium lamp operating at 500 K to be approximately:
>
> (*i*) 0.0025 nm;
>
> (*ii*) 0.005 nm; or
>
> (*iii*) 0.010 nm.

Response

The linewidth for the hollow cathode lamp is about 0.0025 nm, so (*i*) is the correct answer.

This question revises some of the material covered in Part 2 of the Unit. At low operating pressure of a hollow cathode lamp, pressure broadening is negligible. Only Doppler broadening need be considered. Since Doppler broadening is proportional to the square root of the temperature, and the hollow cathode lamp operates at 1/4 of the flame temperature, the linewidth will be:

$$(500/2000)^{1/2} \times 0.005 \text{ nm, or } 0.0025 \text{ nm.}$$

You may check this calculation if you wish using Eq. 2.3:

$$\Delta \lambda_{1/2} = (2/c)4(2RT/A_r)^{1/2} \lambda$$

Thus the linewidth is much narrower for the hollow cathode lamp emission than for flame absorption, satisfying the requirement for the 'lock and key' effect.

****

SAQ 3.2b

Which of the following statements are correct?

(*i*) The measured absorbance in atomic absorption spectroscopy is not dependent on the linewidth of the source.

(*ii*) Hollow cathode lamps are less prone to self-absorption than electrodeless discharge lamps.

(*iii*) The filler gas in hollow cathode lamps can give rise to interferences.

(*iv*) Very high resolution monochromators are necessary for use with hollow cathode lamps.

(*v*) Glass windows are satisfactory for hollow cathode lamps.

Response

Statement (i) is incorrect.

As was shown in the early part of this section, the absorbance is only independent of the source linewidth if this is equal to or less than the absorption linewidth. When the source linewidth is greater than the absorption linewidth, the absorbance will be underestimated.

Statement (ii) is correct.

Self-absorption (or self-reversal) is more likely to be a problem with more intense light sources, since the concentration of atoms will be higher. The exact degree of self-absorption will be determined by the lamp design and operating conditions. In general, however, electrodeless discharge lamps would be expected to be more prone to self-absorption as they are more intense than hollow cathode lamps.

Statement (iii) is correct.

The filler gas will give rise to atomic emission lines, which may be close enough to th emission line of the metal to cause problems in spectrometer opertion and absorbance measurement.

Statement (iv) is incorrect.

Because of the 'lock and key' effect, arising from the exact wavelength coincidence of source emission and sample absorption lines, the monochromator is only required to isolate the spectral line of interest, for which a resolution of 0.1 nm is adequate.

Statement (v) is not generally true.

As discussed in Part 2 of the Unit, glass absorbs over much of the ultra-violet region. For element lines in this region of the spectrum, a quartz window is needed.

****

SAQ 3.3a

Which of the following statements is correct?

(*i*) Gratings have higher resolving power than prisms.

(*ii*) Prisms have a constant (wavelength-independent) dispersion.

(*iii*) Gratings do not give overlapping orders, whereas prisms do give overlapping orders.

(*iv*) Monochromators should contain as few reflecting surfaces as possible.

Response

Statement (i) is correct.

Gratings are capable of very high resolving powers especially for higher orders (see Eq. 3.5; $R = nN$).

Statement (ii) is incorrect.

The dispersion is constant for a grating but varies strongly with wavelength for prisms.

Statement (iii) is incorrect.

Gratings do give overlapping orders but prisms do not. As described in this Section, prisms are sometimes used to isolate a particular order of diffraction in a grating monochromator.

Statement (iv) is correct.

At each reflecting surface there is some loss of intensity. To obtain the best signal to notice ratio, the number of reflecting surfaces should be kept to a minimum.

****

SAQ 4.1a

Which of the following solvents are suitable for solvent extraction from aqueous solutions?

(*i*) Propanone

(*ii*) Methyl benzene

(*iii*) *n*-Hexane

(*iv*) Tetrachloromethane

Response

(*i*) Propanone is unsuitable since it is completely miscible with water.

(*ii*) and (*iv*) Both methyl benzene and tetrachloromethane are immiscible with water and may be used for extraction. They can give rise to errors in the analysis however, due to the associated molecular absorption or emission of the solvent.

(*iii*) Hexane satisfies the requirements of immiscibility with water and does not give rise to problems in the flame, so that hexane is suitable for solvent extraction provided it dissolves the complex more readily than water.

****

SAQ 4.1b

Will higher temperatures increase or decrease the interferences due to:

(*i*) ionisation equilibria, and

(*ii*) association equilibria?

Response

As the temperature is raised the extent of ionisation will increase since ionisation is an endothermic process (le Chatelier's principle), so the interference due to ionisation will become more marked.

On the other hand, molecules are formed exothermically, so that the extent of association will decrease at higher temperatures. Atomisation will be more complete and the interference due to association will be reduced.

****

SAQ 4.1c

How many standards should you use for a calibration curve?

(*i*) 3

(*ii*) 5

(*iii*) 7

Response

We have to bear in mind that the plot will probably not be linear over the whole range of the calibration. With only three points it is not easy to both see where the curvature begins and to obtain a reliable straight line portion of the graph. If one of the standards were made up incorrectly, it may not be obvious which is the poor standard. Although seven points on the curve will give the best representation of the calibration plot, there is a trade-off between analysis time and accuracy and the law of diminishing returns sets in above about five points on the plot. That is, the increased accuracy obtained by using extra standards is minimal and does not justify the extra preparation and analysis time involved.

In practice five standards are usually used, although if you know that none of the samples will yield absorbance readings near the curved part of the plot, then four standards in the linear region may be adequate.

****

SAQ 4.1d

5.0 cm^3 aliquots of waste water were analysed for cadmium by adding various amounts of a standard solution of known cadmium concentration, and making the resulting solution up to a volume of 10.0 cm^3. Use the data below to determine the concentration of cadmium in the original water samples by means of a standard additions plot?

Added Cd /mg dm^{-3}	Absorbance
0.0	0.070
0.2	0.112
0.4	0.156
0.6	0.194

Response

The concentration of cadmium in the waste water is 0.66 mg dm^{-3}.

The standard additions plot of the data is shown in Fig. 4.1e. The intercept on the concentration axis is -0.33. If X is the concentration of cadmium in the original water sample:

$$X/2 = 0.33 \text{ mg dm}^{-3}$$

or

$$X = 0.66 \text{ mg dm}^{-3}$$

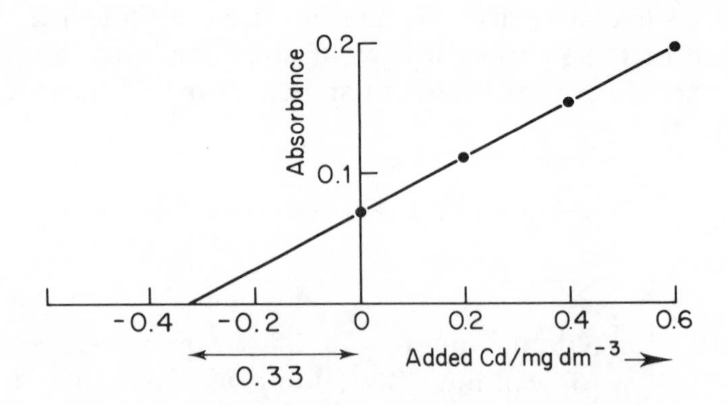

Fig. 4.1e. *Standard additions plot for cadmium analysis*

****

SAQ 4.4a

Which of the four flame regions (1) to (4) shown in Fig. 4.4a is the best region for the analysing light beam in the spectrometer to pass through?

Response

The correct answer is region (3), the inter-conal region.

In region (1) the temperature is too low for atom formation. Also there may be insufficient time for solvent evaporation and sample vaporisation.

The primary reaction zone region (2) is the hottest in the flame, and will therefore give the most efficient atomisation. There are two problems associated with using this region. One is the intense background emission, but the more important limitation is perhaps due to the size of the region. Since it is about 0.1 mm or less thick, the light beam cannot sample the region reproducibly since any flicker in the flame would cause the region to move in and out of the light beam, producing high noise levels.

The secondary reaction zone region (4) is again a region of high background light emission. It is also rather cooler than the inter-conal zone due to the entrainment of cooler air into the flame, so that atomisation will not be as efficient as in regions (2) and (3). We shall also see shortly that there is an increase in oxygen concentration in this region which can be a disadvantage for some elements.

The inter-conal zone then has a high temperature for efficient atomisation, is not too narrow, does not give rise to such strong background light emission and is the most suitable region for analysis.

****

SAQ 4.4b

Can you work out the ratio of free to bound sodium at 2500 K where the dissociation constant is 2×10^{-4}? Take the pressure of chlorine to be 1×10^{-6} atm. Compare your ratio with the value of 5 calculated for 2000 K.

Response

The dissociation constant is

$$K_D = P_{Na} \times P_{Cl}/P_{NaCl}$$

$$P_{Na}/P_{NaCl} = K_D/P_{Cl}$$

and substitute for K_D (2×10^{-4}) and P_{Cl} (1×10^{-6} atm.):

$$P_{Na}/P_{NaCl} = 2 \times 10^{-4}/1 \times 10^{-6}$$

$$= 200$$

Only about 0.5% of the sodium is bound and most of the sodium is present as atoms.

Note that even large changes in the chloride concentration will have little effect on the absorption signal at 2500 K so the interference due to the chloride is negligible. This is in contrast to the strong interference which may be present at lower temperatures.

****

SAQ 4.4c

Which of the following statements are correct?

(*i*) Increasing the size of the droplets formed in the nebuliser will increase the efficiency of atomisation in the flame.

(*ii*) Air-propane flames are more likely to be prone to ionisation interferences than air-acetylene flames.

(*iii*) Nitrous oxide-acetylene flames are less likely to be prone to association interferences (such as oxide or phosphate formation) than air-acetylene flames.

(*iv*) Addition of a potassium salt to a barium solution will reduce the concentration of barium atoms in the flame.

(*v*) Addition of a strontium salt to a calcium solution will reduce the interference by phosphate on the calcium analysis.

(*vi*) Air-acetylene burners can be used for nitrous oxide-acetylene flames.

(*vii*) The sensitivity for the analysis of trace amounts of nickel in aqueous solutions can be enhanced by extracting the nickel into 4-methyl-2-pentanone (MIBK) using ammonium pyrrolidine dithiocarbamate (APDC) as a chelating agent.

Response

Statement (i) is incorrect. Large droplets of solution will evaporate more slowly than small droplets, so the rate and efficiency of atomisation may be reduced for larger droplets.

Statement (ii) is incorrect. Air-propane flames are cooler than air-acetylene flames, and there will be less ionisation in the cooler flame.

Statement (iii) is correct. In the hotter nitrous oxide-acetylene flame, refractory molecules such as oxides will be dissociate to atoms much more rapidly than in the cooler air-acetylene flame.

Statement (iv) is incorrect. Addition of a potassium salt will suppress the ionisation of the barium atoms. The barium atom concentration will be increased.

Statement (v) is correct. Strontium is a protecting agent which will complex the phosphate present, reducing the complex formation between the calcium and the phosphate.

Statement (vi) is emphatically incorrect. The nitrous oxide-acetylene flame has a higher burning velocity than the air-acetylene flame, and may cause 'flash-back' if the area of the burner slot is too large.

Statement (vii) is correct. The extraction step can be used to concentrate the sample by using a much smaller volume of extracting solvent than the volume of the aqueous solution. Also the organic solvent will have a greater efficiency of nebulisation so that more of the sample will enter the flame. The atomisation efficiency may also be greater for the organic solution because of the higher flame temperatures reached.

****

SAQ 5.3a

Would the peak absorbance shown in Fig. 5.3b be increased or decreased by carrying out the atomisation at a lower temperature?

Response

Atomisation must be an activated process since it is so strongly endothermic. The rate of atomisation must then increase with temperature to give a sharper response as shown in Fig. 5.3f. The peak absorbance will be higher at high temperature, although the *area* of the two absorbance-time curves should be the same.

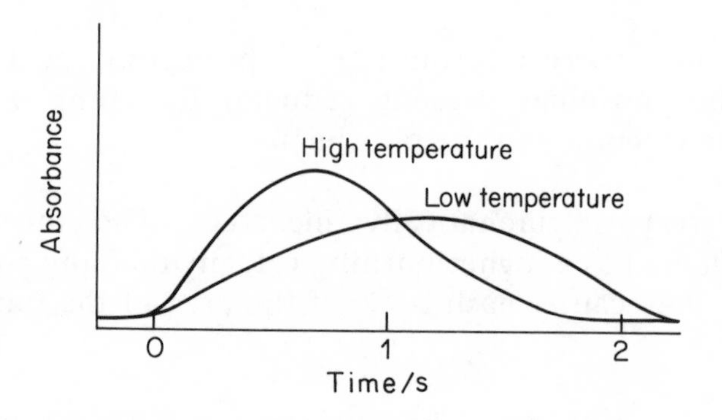

Fig. 5.3f. *Absorbance as a function of time for two different furnace temperatures*

****

SAQ 5.3b

In graphite furnace-AAS the large gas expansion which occurs as the sample volatilises, limits the residence time of the atoms in the atom cell. Some manufacturers use a pressurised cell to keep the atoms in the furnace to increase the residence time and hence sensitivity. This can be done by putting windows at the ends of the furnace, or by flowing gas into the cell from both ends. Will background absorption be more or less of a problem than in the unpressurised cell?

Response

Yes, background absorption will be more of a problem in the pressurised cell. More of the molecules which cause the background absorption will be confined to the furnace. Particulate material which scatters the light beam, increasing the measured absorbance reading will also be trapped in the cell.

****

SAQ 5.3c

Cadmium in water effluent was analysed by the method of standard additions using furnace-AAS. In one experiment a normal furnace was used and the following results obtained:

	Absorbance
sample	0.078
sample + 0.3 ppm Cd	0.140
sample + 0.6 ppm Cd	0.205

It was felt there might be some suppression of signal due to atomisation off the wall occurring before the gas had heated up so the experiment was then repeated using a platform furnace. The following results were obtained: →

SAQ 5.3c (cont.)

	Absorbance
sample	0.090
sample + 0.3 ppm Cd	0.160
sample + 0.6 ppm Cd	0.240

For each set of results calculate the concentration of cadmium in the effluent. Can you say whether the problem of condensation is likely to be present in the normal furnace?

(The method of standard additions was described in Part 4 of the Unit – you should plot absorbance on the y-axis against added concentration of cadmium on the x-axis, and extrapolate back to the x-intercept to obtain the cadmium concentration of the sample. If in doubt check through SAQ 4.1d.)

Response

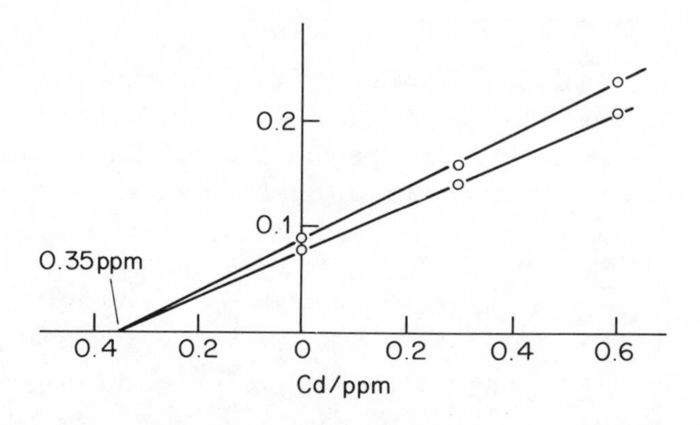

Fig. 5.3g. *Standard additions plot for cadmium samples*

The data are plotted by the standard additions method in Fig. 5.3g. Both sets of data give the same value of 0.35 ppm for the cadmium concentration (not allowing for the dilution factor used in the preparation of the solutions). The plot shows that although the two sets of data give different slopes, they have the same intercept on the concentration axis. The lower slope obtained with the normal furnace is due to the problem of condensation. The implication is that it is essential to use the time-consuming technique of standard additions with the normal furnace, but this may not be necessary for the platform furnace. (To check this a simple calibration curve for the standards should be parallel to the standard addition curve.)

****

SAQ 5.4a

Your laboratory has an atomic absorption spectrometer with facilities for flame, furnace and hydride techniques. Decide which technique you would use for the following measurements.

(*i*) The owner of a fishery is having problems which might be caused by barium leaching from a nearby disused barytes mine into the river upstream of the fishery. Levels of a few mg dm^{-3} of barium are likely to be detrimental to fish.

(*ii*) The waste effluent from a factory which is thought to contain arsenic at a concentration of about 0.1 mg dm^{-3}.

(*iii*) A forensic scientist gives you a few milligrams of plastic for identification. The type of plastic is known but the manufacturer of the plastic needs to be established. One manufacturer uses zinc oxide as a flame retardent, while another uses antimony oxide as the flame retardent. Both oxides would be present at levels of a few percent (w/w) of the plastic.

Response

(*i*) Flame-AAS can be used to determine barium at the level of a few mg dm^{-3}. Since the flame technique can cope satisfactorily with the samples, and there is no problem in obtaining large quantities of the sample, there is no advantage in using the furnace technique, with longer analysis times and poorer precision.

(*ii*) 0.1 mg dm^{-3} is too low for flame-AAS, because arsenic is such a difficult element, with a primary resonance line at 193.7 nm. The hydride method has sufficient sensitivity and would be suitable as arsenic forms a volatile hydride. Furnace-AAS could also be used.

(*iii*) The amounts of zinc or antimony present are likely to be a few tens of micrograms. The simplest and fastest way to test for zinc and antimony at these low levels would be to use a furnace to analyse the solids directly. No pre-treatment is required, although a careful ashing procedure is necessary, since antimony is a relatively volatile element, to decompose the plastic matrix before atomisation, and background correction is likely to be essential as the plastic may not be completely decomposed to volatiles during ashing. High precision is not required in this analysis, as we only need to know which element is present.

****

SAQ 5.5a

0.1 cm^3 of an aqueous solution containing 0.01 mg dm^{-3} of mercury was analysed by the cold vapour method. Can you work out how many atoms of mercury (A_r (Hg) = 200.6) are present?

Response

1000 cm^3 of 0.01 mg dm^{-3} solution contains 0.01 mg of mercury, so 0.1 cm^3 of the solution contains 10^{-6} mg or 10^{-9} g of mercury.

Since the relative atomic mass of mercury is 200.6 , this amount corresponds to $10^{-9}/200.6$ or 5×10^{-12} moles.

Multiplying by the Avagadro constant, the total number of atoms is then $5 \times 10^{-12} \times 6 \times 10^{23}$ or 3×10^{12}.

(Note that if we assume the mercury is flushed into 100 cm^3 of air, the concentration of atoms, 3×10^{10} cm^{-3}, in the gas is a small fraction of the saturation concentration of atoms which is about 3×10^{13} cm^{-3}).

****

SAQ 7.2a

Which of the following statements is correct?

(*i*) Flame emission spectroscopy is not subject to the same matrix interferences as flame-AAS.

(*ii*) Spectral interferences can be reduced by using a smaller spectral bandpass for the monochromator.

(*iii*) 2-point background correction is satisfactory for strongly sloping background emission.

Response

(*i*) *Incorrect*. The same flames are used for absorption and emission so the same matrix interferences will be present.

(*ii*) *Correct*. The smaller the monochromator bandpass the fewer spectral lines will be transmitted.

(*iii*) *Incorrect*. For a strongly sloping background emission it is necessary to average the background signals on either side of the emission line (3-point correction).

****

SAQ 7.6a

Which of the following statements are correct?

(*i*) MIP and DCP methods give lower detection limits than ICP methods.

(*ii*) Flame-AAS is a better method than ICP for the analysis of refractory elements such as aluminium and phosphorus.

(*iii*) A much slower carrier gas flow through the nebuliser is required for an ICP torch than for a air-acetylene flame.

(*iv*) Self-absorption is more of a problem than spectral interferences in ICP spectroscopy.

Response

(*a*) *Incorrect*. ICP detection limits are typically an order of magnitude better than for DCP and very much better than for MIP methods.

(*b*) *Incorrect*. Complete atomisation of refractory elements is very difficult to achieve in flames. The longer residence times and higher temperatures in an ICP ensure virtually complete atomisation.

(*c*) *Correct*. The slower carrier gas flow in the ICP nebuliser requires a different nebuliser design compared to flame nebulisers.

(*d*) *Incorrect*. Self-absorption is insignificant in ICP spectroscopy, while spectral interferences are one of the main problems.

****

SAQ 8.1a

Which is the best method for the following analyses?

(*i*) Aqueous samples of selenium at levels of 10 ng cm^{-3} (ppb).

(*ii*) River water samples for 20 or 30 elements at levels from 0.01 to 100 mg dm^{-3} (ppm).

(*iii*) Sodium in body fluids.

(*iv*) Iron in an organometallic compound (10%).

(*v*) Lead in street dust (0.5%).

(*vi*) Multi-element alloy determination, including phosphorus.

(*vii*) Trace levels of cadmium in food (< 0.1 mg kg^{-1}).

(*viii*) Mercury in fish (< 0.1 mg kg^{-1})

Response

(*i*) At these trace levels, hydride generation is the most sensitive method since selenium is a difficult element to analyse by other methods because of the short wavelength resonance line.

(*ii*) ICP is the most suitable method here, because of its multi-element capacity, high linear dynamic range and relative freedom from interferences.

(*iii*) The simplest and cheapest method for relatively high levels of sodium is flame photometry, although flame-AAS or ICP analysis could be used. The relatively poor sensitivity of the flame photometer is not a problem since the concentration of sodium in body fluids is fairly high.

(*iv*) At these relatively high levels, the simplest and most precise method is flame-AAS.

(*v*) As for (*iv*), flame-AAS is the most suitable method.

(*vi*) If available, a direct reading spectrometer is best. Failing that, ICP would be most suitable.

(*vii*) At trace levels cadmium is best determined by furnace-AAS.

(*viii*) Furnace-AAS is not especially sensitive for mercury analysis (detection limit 0.005 mg dm^{-3}). While furnace-AAS could be used, with rather poor precision, better sensitivity and precision could be obtained with the mercury cold vapour method.

Units of Measurement

For historic reasons a number of different units of measurement have evolved to express quantity of the same thing. In the 1960s, many international scientific bodies recommended the standardisation of names and symbols and the adoption universally of a coherent set of units—the SI units (Système Internationale d'Unités)—based on the definition of five basic units: metre (m); kilogram (kg); second (s); ampere (A); mole (mol); and candela (cd).

The earlier literature references and some of the older text books, naturally use the older units. Even now many practicing scientists have not adopted the SI unit as their working unit. It is therefore necessary to know of the older units and be able to interconvert with SI units.

In this series of texts SI units are used as standard practice. However in areas of activity where their use has not become general practice, eg biologically based laboratories, the earlier defined units are used. This is explained in the study guide to each unit.

Table 1 shows some symbols and abbreviations commonly used in analytical chemistry; Table 2 shows some of the alternative methods for expressing the values of physical quantities and the relationship to the value in SI units.

More details and definition of other units may be found in the *Manual of Symbols and Terminology for Physicochemical Quantities and Units*, Whiffen, 1979, Pergamon Press.

Table 1 *Symbols and Abbreviations Commonly used in Analytical Chemistry*

Symbol	Meaning
Å	Angstrom
$A_r(X)$	relative atomic mass of X
A	ampere
E or U	energy
G	Gibbs free energy (function)
H	enthalpy
J	joule
K	kelvin (273.15 + t °C)
K	equilibrium constant (with subscripts p, c, therm etc.)
K_a, K_b	acid and base ionisation constants
$M_r(X)$	relative molecular mass of X
N	newton (SI unit of force)
P	total pressure
s	standard deviation
T	temperature/K
V	volume
V	volt (J A^{-1} s^{-1})
a, a(A)	activity, activity of A
c	concentration/ mol dm^{-3}
e	electron
g	gramme
i	current
s	second
t	temperature / °C
bp	boiling point
fp	freezing point
mp	melting point
$\approx$	approximately equal to
$<$	less than
$>$	greater than
e, exp(x)	exponential of x
ln x	natural logarithm of x; $\ln x = 2.303 \log x$
log x	common logarithm of x to base 10

Table 2 *Alternative Methods of Expressing Various Physical Quantities*

1. **Mass (SI unit : kg)**

$$g = 10^{-3}\ kg$$
$$mg = 10^{-3}\ g = 10^{-6}\ kg$$
$$\mu g = 10^{-6}\ g = 10^{-9}\ kg$$

2. **Length (SI unit : m)**

$$cm = 10^{-2}\ m$$
$$Å = 10^{-10}\ m$$
$$nm = 10^{-9}\ m = 10Å$$
$$pm = 10^{-12}\ m = 10^{-2}\ Å$$

3. **Volume (SI unit : m^3)**

$$l = dm^3 = 10^{-3}\ m^3$$
$$ml = cm^3 = 10^{-6}\ m^3$$
$$\mu l = 10^{-3}\ cm^3$$

4. **Concentration (SI units : mol m^{-3})**

$$M = mol\ l^{-1} = mol\ dm^{-3} = 10^3\ mol\ m^{-3}$$
$$mg\ l^{-1} = \mu g\ cm^{-3} = ppm = 10^{-3}\ g\ dm^{-3}$$
$$\mu g\ g^{-1} = ppm = 10^{-6}\ g\ g^{-1}$$
$$ng\ cm^{-3} = 10^{-6}\ g\ dm^{-3}$$
$$ng\ dm^{-3} = pg\ cm^{-3}$$
$$pg\ g^{-1} = ppb = 10^{-12}\ g\ g^{-1}$$
$$mg\% = 10^{-2}\ g\ dm^{-3}$$
$$\mu g\% = 10^{-5}\ g\ dm^{-3}$$

5. **Pressure (SI unit : N m^{-2} = kg m^{-1} s^{-2})**

$$Pa = Nm^{-2}$$
$$atmos = 101\ 325\ N\ m^{-2}$$
$$bar = 10^5\ N\ m^{-2}$$
$$torr = mmHg = 133.322\ N\ m^{-2}$$

6. **Energy (SI unit : J = kg m^2 s^{-2})**

$$cal = 4.184\ J$$
$$erg = 10^{-7}\ J$$
$$eV = 1.602 \times 10^{-19}\ J$$

Table 3 *Prefixes for SI Units*

Fraction	Prefix	Symbol
10^{-1}	deci	d
10^{-2}	centi	c
10^{-3}	milli	m
10^{-6}	micro	μ
10^{-9}	nano	n
10^{-12}	pico	p
10^{-15}	femto	f
10^{-18}	atto	a

Multiple	Prefix	Symbol
10	deka	da
10^{2}	hecto	h
10^{3}	kilo	k
10^{6}	mega	M
10^{9}	giga	G
10^{12}	tera	T
10^{15}	peta	P
10^{18}	exa	E

Table 4 *Recommended Values of Physical Constants*

Physical constant	Symbol	Value
acceleration due to gravity	g	$9.81\ \text{m s}^{-2}$
Avogadro constant	N_A	$6.022\ 05 \times 10^{23}\ \text{mol}^{-1}$
Boltzmann constant	k	$1.380\ 66 \times 10^{-23}\ \text{J K}^{-1}$
charge to mass ratio	e/m	$1.758\ 796 \times 10^{11}\ \text{C kg}^{-1}$
electronic charge	e	$1.602\ 19 \times 10^{-19}\ \text{C}$
Faraday constant	F	$9.648\ 46 \times 10^{4}\ \text{C mol}^{-1}$
gas constant	R	$8.314\ \text{J K}^{-1}\ \text{mol}^{-1}$
'ice-point' temperature	T_{ice}	273.150 K exactly
molar volume of ideal gas (stp)	V_m	$2.241\ 38 \times 10^{-2}\ \text{m}^3\ \text{mol}^{-1}$
permittivity of a vacuum	ϵ_0	$8.854\ 188 \times 10^{-12}\ \text{kg}^{-1}\ \text{m}^{-3}\ \text{s}^4\ \text{A}^2\ (\text{F m}^{-1})$
Planck constant	h	$6.626\ 2 \times 10^{-34}\ \text{J s}$
standard atmosphere pressure	p	$101\ 325\ \text{N m}^{-2}$ exactly
atomic mass unit	m_u	$1.660\ 566 \times 10^{-27}\ \text{kg}$
speed of light in a vacuum	c	$2.997\ 925 \times 10^{8}\ \text{m s}^{-1}$